Springer Series on Cultural Computing

Cultural Computing is an exciting, emerging field of Human Computer Interaction, which covers the cultural impact of computing and the technological influences and requirements for the support of cultural innovation. Using support technologies such as artificial intelligence, machine learning, location-based systems, mixed/virtual/augmented reality, cloud computing, pervasive technologies and human-data interaction, researchers can explore the differences across a variety of cultures and cultural production to provide the knowledge and skills necessary to overcome cultural issues and expand human creativity.

This series presents monographs, edited collections and advanced textbooks on the current research and knowledge of a broad range of topics including creativity support systems, creative computing, digital communities, the interactive arts, cultural heritage, digital culture and intercultural collaboration.

This Series is abstracted/indexed in Scopus.

James Hutson

Art and Culture in the Multiverse of Metaverses

Immersion, Presence, and Interactivity in the Digital Age

Springer

James Hutson
Department of Art History, AI and Visual
Culture
College of Arts and Humanities
Lindenwood University
Saint Charles, MO, USA

ISSN 2195-9056 ISSN 2195-9064 (electronic)
Springer Series on Cultural Computing
ISBN 978-3-031-66319-2 ISBN 978-3-031-66320-8 (eBook)
https://doi.org/10.1007/978-3-031-66320-8

This Springer imprint is published by the registered company Springer Nature Switzerland AG
The registered company address is: Gewerbestrasse 11, 6330 Cham, Switzerland

Whatever prestige the bourgeoisie may today be willing to grant to fragmentary or deliberately retrograde artistic tentatives, creation can now be nothing less than a synthesis aiming at the construction of entire atmospheres and styles of life... A unitary urbanism—the synthesis we call for, incorporating arts and technologies—must be created in accordance with new values of life, values which we now need to distinguish and disseminate...

Gil J. Wolman

This book is dedicated to my wife, Piper Hutson, whose unwavering support and unwavering belief in the power of interdisciplinary scholarship have been a constant source of inspiration throughout my academic career. Her intellect, creativity, and passion for knowledge have continually pushed me to explore new avenues of research and to strive for excellence in my work. Piper, your presence in my life is a gift beyond measure, and I am deeply grateful for the love, encouragement, and intellectual companionship you provide.

I would also like to express my heartfelt gratitude to my two children, Bishop and Aurora. Your boundless curiosity, youthful energy, and ability to see the world with fresh eyes have reminded me of the importance of wonder and imagination in the pursuit of knowledge. Thank you for your patience during the countless hours spent writing and researching, and for your constant reminders to approach my work with joy and enthusiasm.

To Piper, Bishop, and Aurora, this book is a testament to the power of interdisciplinary collaboration and the profound impact that supportive and nurturing relationships can have on intellectual pursuits. Your presence in my life has enriched my journey as a scholar and deepened my understanding of the intricate connections between art, culture, and technology. Thank you for being my pillars of strength and for being the driving force behind this endeavor.

Preface

This book serves as a primer for those interested in the world of digital embodiment, where immersive experiences, virtual realities, and interactive environments intersect with the realms of art, culture, and heritage. The primary focus is on the transformative potential of extended reality (XR) and spatial computing technologies in reshaping our understanding of art and culture through an art historical lens. In an era of rapidly advancing technology, we find ourselves at the forefront of a paradigm shift in the study and preservation of cultural heritage. The broader availability and adoption of XR, with its core elements of *immersion, presence*, and *interactivity*, offer new perspectives and avenues for exploring the intricate relationship between location, place, and identity within the domain of art history.

As we embrace this transformative potential, it is essential to define two key concepts: the "metaverse" and "spatial computing." While there may be disagreements in terminology within industry and academia, the underlying concepts remain the same. The metaverse refers to a collective virtual shared space, encompassing all virtual worlds, accessed along the spectrum of XR, and the Internet itself. On the other hand, spatial computing refers to the integration of digital content and tools into the physical environment, creating immersive and natural experiences. Already, we have glimpses of the metaverse through individual portals that provide access to virtual worlds. The ability to seamlessly move between different viewing and experiential modalities is an inevitability on the horizon, bringing us closer to a future where art, culture, and history can be accessed and explored with unprecedented fluidity and depth.

However, with an overwhelming 90% of the world's data being generated in just the past two years—in large part due to advances in artificial intelligence (AI) and machine learning (ML)—the task of organizing and comprehending this deluge of information has become increasingly unmanageable. It is clear that we require an epistemological framework to navigate and extract meaningful insights from the vast amount of data we are generating. As such, this book provides an examination of the critical and methodological underpinnings of the field, establishing the necessary groundwork for a theoretical comprehension of immersive realities within the context of studying and reframing visual culture in the realms of digital art history and digital humanities.

By aligning the elements of immersive realities with essential concepts in art history, we uncover the immense potential of digital embodiment to unveil hidden insights about art and culture spanning across history, simultaneously providing valuable illumination on our own positioning within this ever-evolving narrative. The exploration is, therefore, structured around three central themes: *Location-Immersion, Place-Presence*, and *Identity-Interactivity*. Each paired theme serves as relational understanding to assist in comprehending how these immersive experiences enhance our perception and engagement with art and cultural heritage. By investigating how immersion facilitates the experience of historical locations, how presence illuminates the cultural-historical significance of places, and how interactivity empowers users to enact and explore identity, the treatment calls for a new era of digital art historical research and pedagogy.

Canonical scholarship and contemporary examples are interwoven throughout the book, illustrating the transformative potential of virtual environments and spatial computing in the realm of art and cultural heritage; it showcases how immersive experiences transport users to distant times and places, enabling them to witness art and culture in unprecedented ways. Moreover, the following highlights the ways in which presence bridges the gap between the virtual and physical realms, fostering a profound connection with cultural artifacts.

The potential of interactivity is also explored, as users are empowered to actively engage with and shape their digital environments. Spatial computing, in its essence, grants us the remarkable ability to traverse the dimensions of history, enabling us to seamlessly zoom in and out of the temporal fabric. This unique capability facilitates a simultaneous exploration of the microscopic details and telescopic vistas of the past, unveiling new insights into the intricate shape and dynamics of time. Through the lens of spatial computing, we are empowered to navigate the intricate tapestry of history with unparalleled precision and depth, unraveling hidden narratives and forging connections that would otherwise remain obscured. This transformative capacity offers a profound reimagining of our relationship with time, opening up new avenues for understanding and engaging with the complexities of our collective human experience.

Exploring history through the lens of spatial computing and in the metaverse builds upon the foundational work of George Kubler's *The Shape of Time: Remarks on the History of Things* (1962). Kubler challenged the traditional notion of style by proposing a broader continuum for understanding the evolution of objects and images over time. By embracing the concept of a larger temporal framework, the confines of stylistic categorization are transcended and history as a dynamic and interconnected flow of change can be perceived. In a similar vein, spatial computing empowers us to delve deeper into this continuum, offering a transformative tool to examine the shape of time, uncovering hidden patterns, and shedding new light on the intricate web of historical progression.

Moreover, the treatment embraces the opportunities presented by advancements in artificial intelligence (AI), machine learning (ML), computer vision, and natural language processing (NLP). These technologies have streamlined and expanded the creation and exploration of virtual worlds, making them more accessible than ever to bring historical locations, works of art, and cultural heritage to life. Through such technologies, deteriorating, lost, or stolen artifacts and cultural heritage locations can be reconstituted and repatriated with unprecedented fidelity and speed. Through such recreations, new understandings of the histories of civilizations will emerge, beckoning scholars to reconsider and rewrite history as new insights are gleaned and connections made between isolated timelines and narratives of cultural and technological evolution.

In all, this book lays the foundation for a new understanding of art and culture through the lens of immersive realities. By integrating the core elements of XR with key art historical concepts, it reveals the transformative potential of digital embodiment to uncover hidden insights and reshape our perception of art and culture throughout history. The journey ahead involves a redefinition of the role of the historian and embracing the immersive power of XR and spatial computing, transcending the boundaries of time and space, and charting a course toward a future where art, culture, and history are experienced and understood in novel and profound ways.

Saint Charles, USA James Hutson

Acknowledgments

My profound gratitude extends to Dr. Jason Lively, whose early support and belief in the potential of extended reality (XR) research laid a strong foundation for the work presented in this book. As Dean of the School of Arts, Media, and Communications at Lindenwood University, his visionary leadership has significantly influenced my research trajectory, and I am immensely grateful for his guidance and mentorship.

I am deeply indebted to Dr. Trent Olsen, whose tireless efforts as a collaborator and co-author have enriched the depth and breadth of this manuscript immeasurably. His insightful perspectives, meticulous research, and steadfast dedication to the field of digital art history have been invaluable. I extend my sincere appreciation to Trent for his intellectual rigor and his continual push to expand the boundaries of our knowledge in this exciting domain. Special acknowledgment goes to my co-authors on various projects, including Andrew Wright, Profs. Jeremiah Ratican, Martin Lang, Bryan Robertson, Peter Cotroneo, and Dr. Jason Lively again for their indispensable contributions to this project. Their expertise across game design, the metaverse, AI and gamification, educational technology, and the arts was instrumental to the work at hand.

I also extend heartfelt appreciation to Dr. Kathi Vosevich, Dean of the College of Arts and Humanities at Lindenwood University, for her unwavering support and encouragement throughout the writing process. Her belief in the importance of interdisciplinary research and her commitment to fostering an environment conducive to scholarly pursuits have been vital in bringing this project to fruition. Gratitude is also due to Provost Bethany Alden-Rivers for her administrative guidance and steadfast support. Her commitment to advancing scholarly excellence and her recognition of the transformative potential of XR technology have been truly invaluable.

I wish to acknowledge the myriad researchers and members of the XR community who generously shared their knowledge and insights, enriching the ideas presented in this manuscript. Special thanks to those who championed the establishment of the XR Lab at Lindenwood University, ensuring all students have access to immersive experiences, thus breaking down barriers of space, time, and financial constraints. Their dedication to expanding the horizons of education is truly commendable. Finally, my heartfelt appreciation goes to the staff and colleagues at Lindenwood University who have supported me in myriad ways, from providing resources and research assistance to offering valuable feedback and encouragement. Your contributions have been instrumental in shaping the trajectory of this research, and I am deeply grateful for your unwavering support. Thank you all for your invaluable contributions, guidance, and support. This work would not have been possible without the collective efforts of each and every one of you.

Contents

About the Author

Dr. James Hutson specializes in multidisciplinary research that encompasses artificial intelligence, neurohumanities, neurodiversity, immersive realities, and the gamification of education. Earning a Bachelor of Arts in Art from the University of Tulsa, a Master of Arts in Art History from Southern Methodist University, and a Ph.D. in Art History from the University of Maryland, College Park, he later acquired additional Master's degrees in Leadership and Game Design from Lindenwood University and additional Ph.D. in Artificial Intelligence at Capitol Technology University (2023). Over the span of his academic career since 2006, he has held various pedagogical and administrative positions across five universities, including Chair of Art History, Assistant Dean of Graduate and Online Programs, and most recently, Lead XR Disruptor and Department Head of Art History and Visual Culture. Notably, his scholarly portfolio includes several books on the application of artificial intelligence in education and cultural heritage, such as *Creative Convergence: The AI Renaissance in Art and Design* (2024), as well as numerous articles and case studies.

List of Figures

List of Tables

Part I
Location and Immersion

Chapter 1
Introduction

Abstract The chapter discusses the transformative impact of technologies such as augmented reality (AR), mixed reality (MR), and virtual reality (VR) within the so-called metaverse, highlighting their utility in documenting, preserving, and making globally accessible various cultural heritage sites and artifacts. The author delineates the evolution and current state of XR applications, emphasizing their role in enhancing accessibility to far-flung monuments and museums, and simulating restoration processes of cultural sites. The chapter also considers the implications of these technologies for democratizing access to cultural heritage, thereby fostering a broader understanding and preservation of global art and culture. It argues that through immersive technologies, barriers of space, time, and socio-economic constraints are being overcome, enabling more inclusive participation in cultural dialogue and education. The section establishes a critical framework for understanding how virtual experiences can serve as powerful tools for cultural engagement and preservation in the digital age.

1.1 The Phenomenology and the Experience of Art and Culture

The use of extended reality (XR)—including augmented reality (AR), mixed reality (MR), and virtual reality (VR)—in the so-called metaverse to experience cultural heritage has greatly expanded over the past two decades. Whether using XR and digital twin technologies to record and preserve sites from antiquity, using head-mounted displays (HMDs) and virtual reality (VR) to study and improve the accessibility of far-flung monuments and museums around the globe, or to simulate the restoration process of cultural heritage sites and artifacts, the use cases for how the metaverse will be leveraged for experiencing global art and culture continues to rapidly grow. At the same time, the full potential of emerging and immersive technologies has yet to be realized. From early 3D models reconstructing ancient sites in Greece and Rome, as seen in Fig. 1.1, to the artificial intelligence (AI)

J. Hutson, *Art and Culture in the Multiverse of Metaverses*, Springer Series on Cultural Computing, https://doi.org/10.1007/978-3-031-66320-8_1

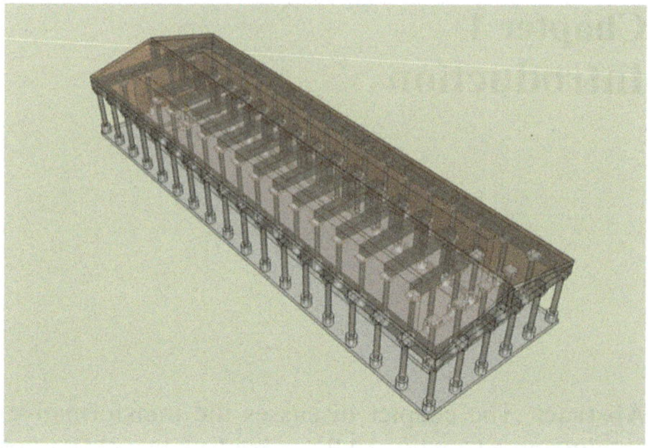

Fig. 1.1 Schematic recreation of second Heraion of Samos, 2016 (from Wikimedia Commons, licensed under CC-BY 4.0)

supported generative 360 content, the potential to be immersed and interact with existing, reconstructed, or imaginary scenes, objects, and persons has never been more visceral.

Moreover, the latest generation of hardware and software now enables the process of bringing touch, temporal properties, taste, and smell to these virtual experiences. The new technologies that are used separately, or are in their infancy, such as the latest haptics (which allow users to feel pressure, texture, and heat and cold) and olfactory VR (OVR), as well as polygonal rendering and low latency of the latest generation of headsets and game engines, allow for ultra-high-resolution presentation? For a more authentic experience of what is or was (Bender and Broderick 2021). The ability now exists to retrace historical occurrences and have viewers inhabit that space to see how events actually unfolded. Moving beyond the limitations of *mise en scène*, an even more immersive experience, complete with feelings of texture, thermal properties, acoustic, and olfactory information specific to a time period offers a more complete view and understanding of cultural heritage.

The availability and accessibility of global cultural heritage as a public good is a priority that extends beyond the limitations of national status, class, gender, race, or ethnicity. Objects such as paintings like Leonardo da Vinci's *Mona Lisa*, Fig. 1.2, monuments like the Pyramids of Giza in Egypt (Fig. 1.3), and places like the Great Barrier Reef in Australia (Fig. 1.4), are collectively held to be important to preserve as these in many ways embody the cultures that created, experienced, and used them. The United Nations Educational, Scientific, and Cultural Organization (UNESCO) codified the "exceptional universal value" of such cultural heritage sites at the 1972 Convention Concerning the Protection of the World Cultural and Natural Heritage and established the idea that a common cultural and national heritage exists that is of equal importance to all of humanity (Gruber 2019).

Access to such cultural heritage is necessary not only for global citizenship and diversity, but also constitutes "cultural potential" for contemporary societies (Pownall et al. 2022). The cultural inheritance from previous generations is our collective world heritage, which also informs humanity's identity in the present and is passed on to

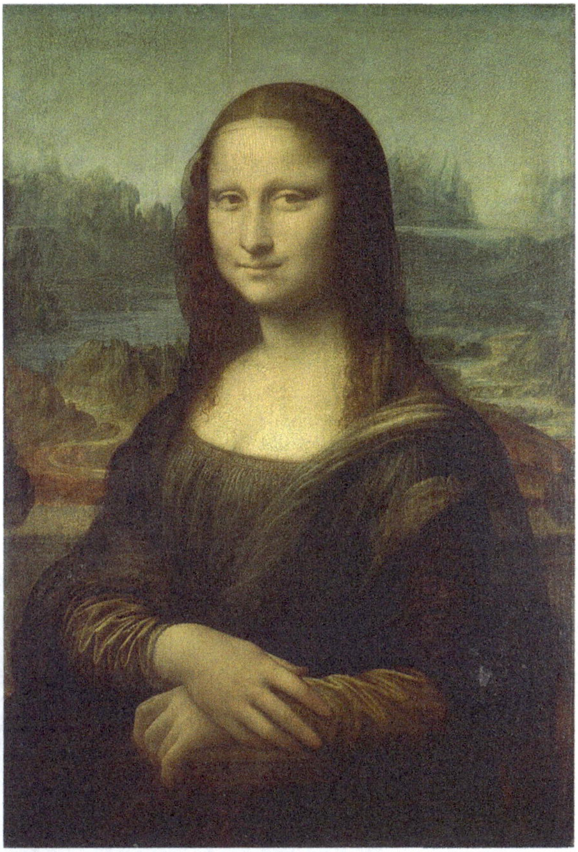

Fig. 1.2 Leonardo da Vinci, Mona Lisa, ca. 1503–06 (from Wikimedia Commons, licensed under CC0)

future generations. Culture itself encompasses the social norms, behaviors, and institutions of a given group. These regional characteristics can include customs, knowledge, and beliefs, as well as rituals, art, and habits. Understanding other cultures than one's own has demonstrable benefits, such as improving empathy, innovation, and collaboration (O'Connell et al. 2022). Given that objects, monuments, and places are often tied to cultural identity, one way to promote an understanding of the "Other" has been to see and/or experience the physical and aesthetic products of that civilization (Maurice 2022). Such cultural heritage also allows for a continuous reevaluation of cultural identity, which in turn transmits skills, memories, experiences, and knowledge from one generation to the next. Exposure to heritage also inspires innovation and creativity in the form of future cultural products. Finally, experiencing cultural heritage also contributes to the enrichment of social capital and creates a sense of individual and collective belonging, promoting social cohesion, peace, and prosperity (Ruiz-Ruiz et al. 2022).

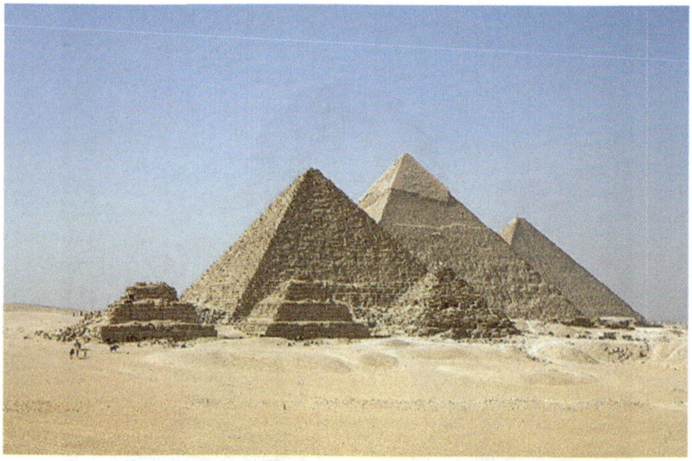

Fig. 1.3 Pyramids of Giza, ca. 2500 B.C.E. (from Wikimedia Commons, licensed under CC0)

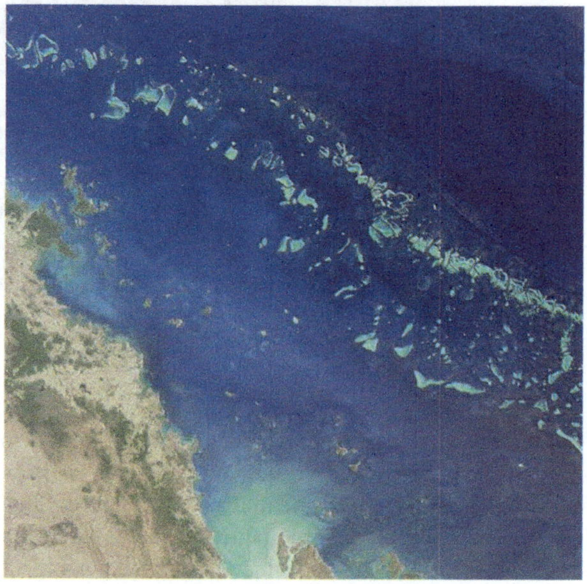

Fig. 1.4 The Great Barrier Reef near Airlie Beach, Whitsunday Islands, Queensland, Australia (from Wikimedia Commons, licensed under CC0)

At the same time, culture is not restricted to the static and stationary remnants of material and visual culture handed down from a given people. Culture is dynamic and contextual; a continuous dialogical relationship that is negotiated between those living in a society and the products, signs, and symbols that help define their belief systems and worldview. Therefore, culture must be experienced to be comprehensible and understood. Such experience is not limited to works of art, monuments, or cities; inherited living expressions are equally important. These expressions include social manners, oral traditions, rituals, performing arts, festivals, ephemeral performances, and techniques for traditional crafts, which all coalesce in practice to enact culture. Thus, ephemeral and intangible cultural heritage, otherwise known as living heritage, is as significant for understanding cultural heritage as monuments carved in stone (Quang 2022).

Despite the research on the benefits of diverse societies and the importance of cultural heritage, the experience of global art and cultural heritage largely remains limited to a small percentage of the population that have the resources to physically visit locations in person (Zhao et al. 2022). Otherwise, the general population had previously to be content with translation and transmission of such experiences via photographic reproductions or videos of distant locations. However, the technological advances of the past few years have resulted in a new portal to access and engage with the world and culture throughout all time and history using tools like virtual reality (VR) and head-mounted displays (HMDs)—a portal popularly known as "The Metaverse" (Ball 2022).

1.2 The Metaverse and Visual Cultural Experiences

The notion of an alternative, "virtual" reality (VR) arose from science fiction, rather than science in the modern era. Examples of the technology appear in *Pygmalion's Spectacles* (1935), *Typewriter in the Sky* (1940), *I (Professor Corcoran)* (1961), *The Three Stigmata of Palmer Eldritch* (1965), and *Neuromancer* (1984). Virtual reality has many definitions, but for our purposes we will rely on the understanding provided by Pope that it is "an experience that encompasses most of the senses, including sight, hearing, and touch, and represents an alternative to reality" (Pope 2018). The term "metaverse" was first coined by Neal Stephenson in his 1992 science fiction novel *Snow Crash*. In the book, the metaverse is a virtual reality-based successor to the Internet, where users can interact with each other and digital objects in a shared space (Tally 2022). Aside from literature, cinema was another inspiration for VR, such as in the 1950s when filmmaker Morton Heilig envisaged an "Experience Theatre" that encompassed the senses beyond flat screens and resulted in the creation of a mechanical VR engine (Milz 2019). The final "Sensorama" in 1962 was designed to engage many senses, including sight, sound, smell, and touch. Following the foray into multisensory experiences, Ivan Sutherland and Bob Sproull created the first head-mounted display (HMD) in 1968 (Champion and Foka 2020). While originally restricted to use in military and medical research communities, the technology and interest in creating a virtual multisensory experience has gone mainstream and become more widely available in recent decades.

With films like *Ready Player One* (2018), immersive virtual environments that have collectively been known as the metaverse are at the forefront of popular imagination, and seemingly within reach. However, as Ball and others define the metaverse as a future interoperable and scalable real-time network that can be experienced in 3D virtually and synchronously, there are further developments with hardware, software, and infrastructure that need to be developed. In the new virtual world, there would be no limit on the number of users and interactions, including social, financial, cultural, and so on, all taking place within this persistent virtual community. The version of the metaverse that we currently see with companies like Meta (shortly to be joined by Google and Apple) is neither interoperable nor scalable. Instead of being inside a virtual or 3D version of the Internet, there are many portals to enter different virtual spaces on different proprietary platforms. Therefore, we actually have a "multiverse of metaverses" existing simultaneously (Valaskova et al. 2022). Each virtual space is uniquely designed for a specific use and purpose with their own avatars for a user's digital embodiment in the experience. Unfortunately, these spaces and avatars are not traversable or transferable across spaces.

Yet, even with these existing limitations, the current version of the metaverse already has the potential to provide a virtual shared space where users can experience art and culture in a simulated world. Such experiences have been developed through digital cultural heritage preservation in digital art history and digital humanities (Hutson and Olsen 2022). These developments began in 2001 with the digitizing of cultural heritage sites. Both ancient and modern sites were created, though access and use were often restricted to researchers and institutions. Such virtual environments (VE) were first developed using a cave automatic virtual environment (commonly referred to as CAVE), an immersive virtual reality setting that utilizes projectors directed toward three to six walls of a room-sized cube. The name itself draws inspiration from Plato's allegory of the Cave in his work *The Republic*, wherein a philosopher contemplates the nature of perception, reality, and illusion. An early example of CAVE technology can be seen in the Foundation of the Hellenic World (FHW) (Fig. 1.5). Digital reconstructions were possible through the use of the technology, beginning with Miletus, an Athenian and later a Roman colony on the coast of Asia Minor (Tzortzaki 2001). The potential was recognized for museums and other cultural heritage institutions almost immediately. For instance, Roussou published on how the technology would be ideally suited for exhibitions in museums as "edutainment" (Roussou 2001). Heeding the call, many museums began integrating extended reality (XR) experiences in the first decade of the twenty-first century, including The Museum of Pure Form and The Virtual Museum of Sculpture. These experiences were designed for use by the general public, who had no experience using this new technology. Unlike experiences for the scientific and medical communities, which tended to be longer, these were necessarily short to keep visitors moving through exhibition spaces. These site-specific XR experiences expanded with examples such as The Exploratorium, a public science museum, and The CREATE project, an EU-funded project that allows users to reconstruct archeological sites.

Around the same time, whole museum collections were digitized for viewing using augmented reality (AR) devices or head-mounted displays (HMD). Beginning in 2006, the Center for the Art of East Asia in the Department of Art History, Division of Humanities, at the University of Chicago (CAEA) started digitizing and archiving their collections of East Asian paintings and sculptures. Such examples of cultural heritage institutions expanding access to art historical resources are driving digital twins today. These virtual representations or reconstructions of a cultural artifact, site,

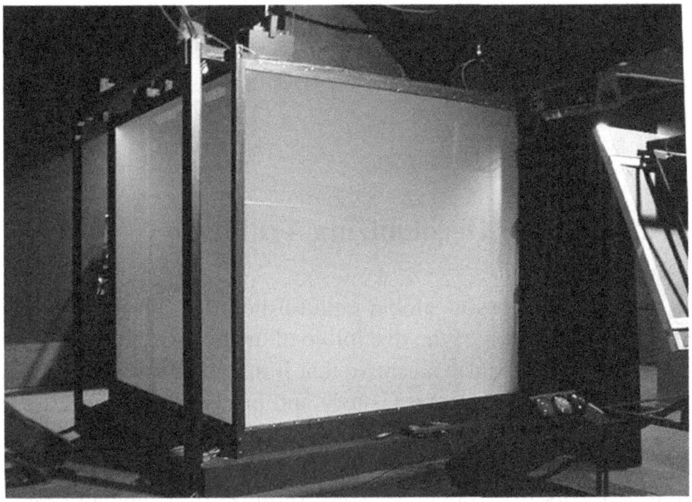

Fig. 1.5 Rear view of the REACTOR, CAVE virtual reality system, Foundation of the Hellenic World (FHW), History Museum, Athens, Greece. Photograph by Dave Pape, May 28, 2001 (from Wikimedia Commons, licensed under CC0)

or heritage asset involves creating a digital counterpart that accurately captures the physical attributes, characteristics, and historical significance of the original object or site. Examples can include 3D models, digital reconstructions, or immersive virtual environments that provide a rich and interactive experience for studying, preserving, and sharing cultural heritage (Luther et al. 2023). Digital twins in cultural heritage enable virtual exploration, analysis, and interpretation of artifacts and sites, facilitating research, education, and public engagement with cultural heritage in a digital format. The digitized collection at CAEA and others move the experience from a physical museum to a completely virtual learning environment (VLE). The development is made possible through the recent availability of user-friendly interfaces and immersive design, coupled with shorter engagement duration, leading to the design of VLE that is currently popular with museums and other cultural heritage institutions.

However, only in the last decade has the use of VLE to deliver cultural heritage content become widely available. Most notably this has been accomplished through the creation of computer-generated museums using 3D models or through digitization efforts of real museum spaces. A watershed moment occurred with the launch of Google's Arts and Culture in 2011, which allowed virtual visits to museums, accessible to anyone with a smartphone or computer. The company seized upon the democratizing potential of the technology with Google Cardboard in 2014, an inexpensive HMD that was quickly adopted for secondary education. Similar VLEs have been developed to tour real or virtual museums (VM), such as the Rijksmuseum, Amsterdam, the National Archeological Museum of Marche in Ancona, and Gyeongju VR Museum, South Korea (Clini et al. 2018; Favro 2006). The Louvre also recently (2021) released their digital collection of nearly half a million works from its collections and made them available online. One of the most impactful projects as of late (overseen by UNESCO, 2017–2020) is the creation of full virtual tours of World Cultural Heritage Sites.

1.3 Decentering and Decolonizing Art History

The efforts to expand access to global cultural heritage developed alongside and supported the decolonization of the discipline of art history itself. The reconsideration of the dominant Eurocentric narrative that frames discussions and valuations of art and culture began with New Art History and has been underway for more than half a century. Even with a clear focus on decentering the West and providing a more equitable, global perspective of the contributions of non-Western arts and cultures, there is still much work to be done. Reasons for the entrenchment of Western hegemony that holds up for admiration of the accomplishments and creative activities of Eurocentric peoples can be found in historiography and the very construct and practice of history itself.

Selective inclusion of events, influential persons, nations, and peoples make up our understanding of who we are as a species. Our understanding of our own place in the world is dependent upon the epistemological investments of different disciplines and the choices (conscious or not) of what to include and what to exclude. Since the Greek historian-geographer Herodotus (ca. 484–425 B.C.E.) invented "history" in the fifth-century B.C.E. to tout the victory of the Greeks over the Persians, the field has been used by the West to track the ascendency of European patriarchal civilization in a heroic narrative. From the ancient Greek term *ίστορία*, meaning the act of inquiry, or seeking knowledge based on investigation, historical methodologies have supported various Eurocentric narratives without alternative constructs to challenge them (Easterling et al. 2006).

While illuminated manuscripts arose in late antiquity, most readers would have experienced such reconstructions of historical events (as real or imagined) solely through the written word on a scroll or codex. The interim between the fall of the Roman Empire and the rise of the Renaissance in Europe witnessed a rich array of illustrations to communicate the significance of historical events. But it would not be until the early modern period that the central place of art—as restricted to the mediums of painting, sculpture, and architecture—would be codified. Born of the Renaissance-era boast of the triumphant return of the glory of ancient Rome after the darkness of the medieval period, the Italian artist-biographer Giorgio Vasari (1511–1574) would craft a compelling narrative that would form the epistemological underpinning of the discipline of the "rebirth" of art and culture in his own time. In his work, Vasari begins by acknowledging the greatness of ancient Roman art as the pinnacle of perfection. He then proceeds to divide the era of "modern" painting, which marks the revival of that perfection, into three distinct periods. According to Vasari, the first period, characterized by the influence of Giotto (d. 1337), witnessed a departure from the *maniera greca*, or "Greek manner," and a renewed focus on the faithful representation of Nature. The second period, led by Masaccio (1401–1428), Donatello (1386–1466), and Brunelleschi (1377–1446), embraced mathematical and scientific approaches, such as linear perspective, which paved the way for the third and final period. In this third period, Vasari identifies Michelangelo Buonarotti (1475–1564) as the preeminent figure, "who transcends and surpasses them all, as he reigns supreme not merely in one of these arts but in all three together [painting, sculpture, and architecture]" (Barocchi 1971, 4:10).

After Vasari's well-crafted rhetorical argument that drew on biographical and biological metaphors from ancient philosophers, the discipline that places aesthetic objects and visual expressions in an historical and stylistic context would be codified for a broader academic audience through the insights of the German art historian and archeologist Johann Joachim Winckelmann (1717–1768) (Winckelmann 1987). The narrative of a cyclical, biological model for understanding the development of each period and phase of Western art and culture was institutionalized within academia in the nineteenth century.

A philosophy of art, based on Vasari and Winckelmann's biological model, was espoused at the outset of the nineteenth century by Georg Wilhelm Friedrich Hegel (1770–1831), who introduced a new evolutionary system in *The Phenomenology of Spirit* in 1807 (2018). Hegel explores the evolution and revelation of knowledge and consciousness throughout history and views art as an expression of the Spirit of cultures and individuals, presenting three main stages of art: symbolic, classical, and romantic. In the symbolic stage, powerful ideas are expressed through various forms that may lead to distortions. Classical art embodies perfect, idealized human forms without distortion, while romantic art emphasizes inwardness and the inadequacy of images to convey deeper ideas. Hegel's claim that art comes to an end has been widely discussed since publication, suggesting that as Spirit reaches full self-realization, the need for physical expressions through art diminishes.

The systematic classification of various stylistic trends over the course of history would find resonance in the academy via the efforts of Austrian art historian Alois Riegl (1858–1905) and Swiss art historian Heinrich Wölfflin (1864–1945) (Riegl 1985; Wölfflin 2009). Through their collective work, the musings of connoisseurs and collectors were made "scientific" and provided a compelling model for understanding all products of human culture as reflective of their broader belief systems (*Weltanschauung*), bound to geographical, chronological, and stylistic conventions, and provided a systematic classification system to discuss art and other forms of cultural production. Alongside the disciplines of sociology, history, and anthropology, art history sought to investigate and better understand past and present cultures through their visual and material culture.

Even with the shift in focus seen with the advent of New Art History, which sought to move beyond the formalist and positivist framework for the discipline by embracing Marxism, semiotics, and deconstruction, the canon of art has proven difficult to disrupt and decenter due to the persuasive narrative constructed over the course of two and a half millennia (Harris 2002). Mid-twentieth-century poststructuralist discourse first attempted to challenge the dominant narrative and called for a critical reexamination of previous assertions, especially in light of the events of World War II. The previous clashes over ownership of the discipline, famously played out in the literary exchanges between the Jewish art historian Meyer Schapiro (1904–1996) and the German nationalist-philosopher Martin Heidegger (1889–1976), would coalesce in an existentialist crisis for the discipline (Heidegger 1971; Schapiro 1968).

Jacques Derrida (1930–2004) would cast such squabbles in a childish light and (in true deconstructivist style) call into question the assumptions the discipline previously had with terms such as "frame," "vision," and "perspective" (Derrida 1987). Despite the flippant nature of his polylogues, Derridian critical strategies of vision and visuality are now more central to the discussion of the metaverse and the metaphysical than ever. The framework provided by Stephen Melville's *The Temptation of New Perspectives* (1990) makes the adoption of literary deconstruction to the study of visual art relevant. Calling into question how we are framing "perspective" and what is "inside" and "outside" of a work of art has led to postcolonial calls to decenter the West and look beyond the narrow confines of Europe and North America in the history of art.

Since then, there have been many attempts to rewrite more inclusive global art history, while grappling with the epistemological ghosts and limitations of scant surviving material and visual culture and primary sources from non-Western cultures (Joyeux-Prunel 2019). Regardless, progress has been made to include more of a global perspective. The first development has been the broader recognition and inclusion of non-Western art traditions—African, Asian, and pre-Columbian American art (Bose 2009). Additionally, there has been a growing focus on transnational and cross-cultural exchange in art history. Scholars have begun to study how different cultures have influenced and interacted with one another throughout history, and how art has been used as a medium for these exchanges, which has led to a more nuanced understanding of the relationships between different cultures and their art (Hager 2020). Finally, the increased attention given to the role of colonialism and imperialism in shaping the art and art history of different cultures has resulted in a rich body of scholarship in Postcolonialism (Amselle et al. 2003). Scholars have begun to examine how the art of colonized peoples was often suppressed or co-opted by colonial powers, and how this has affected the way that art is understood and appreciated today.

1.4 Sight, Vision, and Description: The Unspoken Language of Visual Experience

These new critical and methodological approaches have expanded the scope of the discipline and a global understanding of cultural development. But we are still limited by what we can see, which inadvertently dictates what we say about art and culture. For instance, as Baxandall (1985, p. 1) clarified, "We do not explain pictures: we explain remarks about pictures." In other words, what we can see is directly related to what we can say, and vice versa. As such, the history of visual and ekphrastic description developed alongside technologies to create and disseminate works of art and the visual vocabulary created to engage with it. In antiquity, with limited travel and visual reproductive capabilities, authors would assume the reader likely would not have access to a work being described. As such, when fourth-century C.E. Greek Libanius described a picture in the Council House at Antioch, an ekphrastic description was used that sought to present the representation as a real scene: "Four men were running out of the houses, one of them calling to a lad standing near—for his right hand showed this, as if giving some instructions. Another man was turned towards the first one, as if listening to the voice of a chief. A fourth, coming a little forward from the door, holding his right hand out and carrying a stick in the other, appeared to shout something to other men toiling about a wagon …" (Baxandall 1985, p. 2). The action unfolds as though described by a narrator of an epic tale.

Such an approach was also adopted by Giorgio Vasari for his renaissance biographies with the assumption that his audience may have a passing knowledge of the works he describes, but not reproductions of them for reference. At the same time, the expansion of available prints, starting with woodcuts of the fifteenth century, afforded black-and-white linear reproductions of visual objects, which helped the field evolve a more critical vocabulary. In the nineteenth century, lithography was developed and allowed for more detailed and accurate reproductions of images widely used for commercial printing and illustrations in books and newspapers. While printmaking was instrumental in the spread of artistic and scientific knowledge during the Enlightenment, the advent of photography revolutionized visual reproduction. Starting with early experiments in 1822, the photomechanical process for reproducing imagery allowed for the creation of highly detailed and accurate reproductions of reality and was quickly adopted in many fields. Wölfflin further refined the descriptive capabilities of scholars through his use of photographic techniques and the "magic lantern," seen in Fig. 1.6, which were adopted in universities in the nineteenth century, replacing engraved works of art. The projection of two images side-by-side allowed for more refined, critical comparisons. However, the adoption of projectors was slow due to their perceived use primarily as a device for entertainment (Kohl 2012).

While various forms of printmaking and photography existed in the nineteenth century, color processes would be developed in *fin de siècle*. New technologies, including the halftone process and color lithography, were developed, which allowed for the reproduction of photographs and color images in print. This further expanded the possibilities of visual reproduction and had a profound impact on the field of commercial printing and the study of visual culture. Yet, even as the availability of color reproductions expanded post-World War II, opening up new avenues for study and experience of cultural heritage around the world, these would be limited to two dimensions in low resolution. One could argue that we have been stuck here until quite recently. Only at the end of the last millennium did digital technologies, including the personal computer and the Internet, develop to significantly impact visual reproduction, readily allowing for digital manipulation, storage, and distribution of a variety of imagery in high resolution. This has led to the development of new forms of visual media, such as video, animation, and virtual reality. In the past

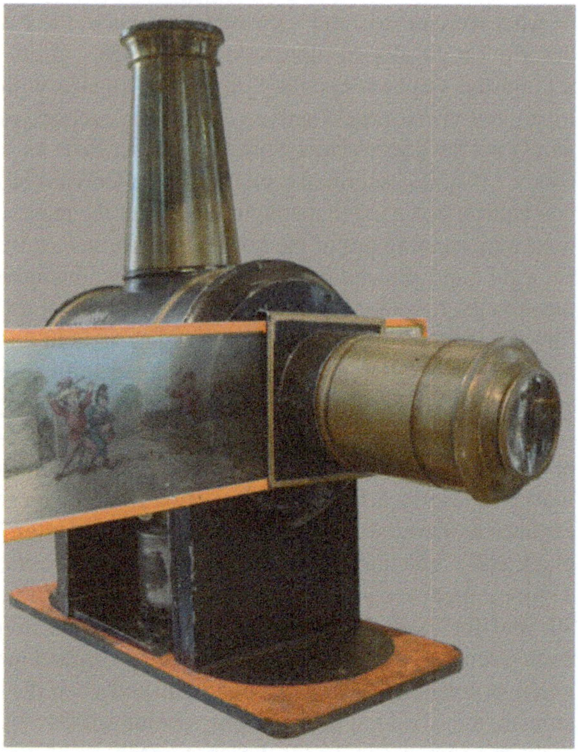

Fig. 1.6 Magic Lantern, Schlossmuseum, Aulendorf. Photo by Andreas Praefcke, March 2006 (from Wikimedia Commons, licensed under CC-BY 4.0)

few years, the general population has finally been granted access to 3D modeling software, 3D printers, and new hardware that are both user-friendly and affordable. With the third dimension digitally conquered, we look ahead to the fourth.

1.5 Location-Presence, Place-Immersion, and Identity-Interactivity: Experiential Dyads

XR devices allow the user to collapse space and time and bring together distant geographical locations across the expanse of history. The hardware and software to support such experiences are becoming increasingly accessible and affordable. The first VR headset that was released to the consumer market that could be considered successful was the Oculus Rift in 2013. While the unit was primarily designed for the gaming market, it also supports educational applications. Since 2013, a number of other VR headsets have been released, representing a range of quality and immersion. For example, the Google Cardboard uses lenses and a smartphone to give users a virtual experience, which is of lower quality and immersion, but does not require expensive hardware or a dedicated physical space. In 2016, the HTC Vive was released and became popular due to its sophisticated optics and its use of the already-popular game distribution service Steam. However, it requires connection to a high-powered computer to run effectively and avoid lagging. Oculus, which was purchased in 2014 by Facebook, began releasing self-contained headsets, such as the Oculus Go and Quest. These VR headsets provide a less sophisticated immersion experience, but are cheaper and do not require an external computer. The usefulness of the technology was further expanded when Facebook released the Oculus Quest 2 in October 2020, followed by the Meta Quest 2 Pro in 2022. While Meta primarily targeted the gaming features of the Quest all-in-one headset, it was also promoted as a work aid in design-related fields, such as architecture and design. Most recently, the mixed-reality (MR) Apple Vision Pro, announced in June 2023, sought to further expand the user base for the technology with their approach to "spatial computing." While the term has only recently come into the popular lexicon, the term spatial computing was coined by MIT researcher Simon Greenwold in 2003 and is a comprehensive term that encompasses a range of technologies aimed at bridging the gap between the digital and physical realms. It focuses on empowering human–computer interactions by transforming the way digital content and tools are interacted with, creating immersive and natural experiences (Bhatt and Freksa 2015). Through the seamless integration of user interfaces into the physical environment, spatial computing utilizes technologies such as AR, VR, MR, and sensors to facilitate three-dimensional perception, natural interaction, and enhanced spatial experiences. Connected to their other devices, the Vision Pro is projected to further broaden the appeal by improving human–computer interfaces and making immersive experiences more intuitive and seamless.

The virtual worlds evolving within the metaverse, and now with spatial computing, are made possible through these ever-more-powerful XR devices, but their use cases differ by type. When considering how XR may be used in cultural heritage research, study, or preservation, the three types must be considered separately (Table 1.1). For instance, augmented reality, or AR, is ideally suited to engaging with an object or place that does (or did) exist but is now ruined or missing. An example may be one of the Seven Wonders of the Ancient World that has since been destroyed like the Colossus of Rhodes or a monument that has only been partially destroyed by earthquakes and human scavenging, such as with the top level of the Colosseum. In these use cases, the viewer needs to see what something looked like in situ or fully reconstituted on location.

On the other hand, unlike AR, mixed reality, or MR, merges the real world with a digitally created one. While the geometry of a space occupied by a user is retained, computer-generated imagery can be overlaid to integrate virtual content seamlessly within the area. As such, MR is ideal for use with architectural plans or structures for cooperative analysis, such as viewing a reconstruction of Angkor Wat in Cambodia. Finally, should the subject of study be geographically distant or in an imaginary location, such as walking within the Great Pyramid at Giza, VR is the ideal solution.

The way in which these experiences of digital embodiment effectively transport viewers is of special interest to those involved in the fields of art and culture. The three factors that make the experiences outlined above possible are the feelings

Table 1.1 Types and use cases for different types of XR

Type	Use case	Example
Augmented reality (AR): overlays digital/ computer-generated content onto the real world	Reconstruct was no longer exists	Providing a digital overlay on-site of the Bamiyan Buddha statues destroyed in 2001 by the Taliban
Mixed reality (MR): merging real-world and digital/ computer-generated one	Reconstitute an entire space/ location with geospatial data retained	Repatriating and reconstituting the Louvre prior to the French Revolution with all decorations and trappings while on-site
Virtual reality (VR): complete sensory immersion in a digital/ computer-generated environment	Transport users to distant/ inaccessible locations	Transport viewer inside the Tomb of Seti I for experience and to preserve the integrity of the site

of *Immersion*, *Presence*, and *Interactivity*, which are ideal to visualize the world at human scale and provide the opportunity to understand the world in such a way that is otherwise impossible. These qualities are most associated with VR and will constitute the focus of this treatment and potential use in the field moving forward. In terms of VR, *immersion* refers to the condition where a user loses the distinction between the real and virtual worlds. Immersion can be enhanced by using high-fidelity visual processing, low latency, realistic sounds, olfactory stimulation (OVR), and haptic feedback, which can make the virtual environment more believable and realistic (Melo et al. 2022). Notably, immersion and the immersive quality of an experience are generally considered in terms of the technical aspects involved in the simulation.

Presence refers to the ability of a user to feel physically present in a virtual location and environment. As opposed to immersion, which is associated with the technical aspects of the virtual system, presence considers the quality and nature of feeling. High levels of presence can make the user feel as if they are actually in the virtual world, while low levels of presence may make the user feel as if they are simply viewing a 3D animation. Presence can be enhanced by providing a sense of agency, or the ability to control the virtual environment (Gibbs et al. 2022). This principle can be illustrated in degrees of freedom. For instance, if experiences are limited to three degrees of freedom (3DoF), and only allow the viewer to move their viewing space and angle via yaw, pitch, and roll, then there will be less of a sense of presence. The effect feels as though one's head is anchored to a tripod, which reflects how the images were captured. On the other hand, six degrees of freedom (6DoF) also allows experiences where the user can move their head in all directions and also move back, forward, up and down in space, which improves feelings of presence.

Interactivity refers to the degree to which the user can interact with and influence the virtual environment. High levels of interactivity allow the user to actively engage with the virtual environment, while low levels of interactivity may limit the user to simply viewing the virtual environment. See the differences in 3DoF and 6DoF above. Building on what improves presence, interactivity can be enhanced by providing the user with a variety of tools and controls, such as hand-held controllers or gesture recognition, that allow them to interact with the virtual environment in a natural and intuitive way (Leung et al. 2022). In combination with presence and immersion, interactivity can contribute to a more lifelike experience which may include haptics or sensory feedback upon interacting with objects in the environment (e.g., bumping into walls or weight and texture of objects picked up and held).

Therefore, *immersion*, *presence*, and *interactivity* are factors that work in tandem to create an immersive and believable experience for users. These considerations are the most often studied in XR research to help identify and quantify the level of each user's experience (Han et al. 2022). If we turn our attention to cultural heritage, however, additional levels of critical and methodological inquiry can be applied to understand "meaning making" with the art historical concepts of *Location*, *Place*, and *Identity*. The three concepts are often used together when analyzing and interpreting works of art and cultural heritage. *Location* refers to the physical location where a work of art was created or is currently located, which can include an artist's *atelier*, workshop or studio, a specific site or building, or a geographic region. The location of a work of art can have a significant impact on its meaning and significance and can provide insight into the social and cultural context in which it was created (Newton and Spencer 1982).

Place, on the other hand, moves beyond geographical location and instead refers to the symbolic or cultural significance of a location. This concept is intimately linked with the notion of a "sense of place," which encompasses the emotional and cultural associations that a specific geographic location evokes. When considering this concept, one can turn to the Parthenon Museum in Athens as an illustrative example. While the majority of the Parthenon marbles are currently housed in London, the Athens museum features an exhibition space where empty placeholders stand in for the absent marbles. Moreover, the museum's expansive glass windows offer panoramic views of the Acropolis. These deliberate design choices serve to emphasize the absence of the Greek marbles from their place of origin, drawing attention to the fact that they have been removed from their homeland (Dear et al. 2011).

Finally, *Identity* refers to the ways in which the individual or collective identity of the artist or viewer is represented or constructed in a work of art, or through the process of viewing or interacting with it. This can include the artist's personal experiences, beliefs, and cultural background, as well as the identities of the subjects depicted in the work of art. Identity can also refer to how the work of art itself contributes to the construction of the artist's or viewer's identity (Soussloff 1999). These three concepts are closely interconnected and can provide insight into the social, cultural, and historical context of a work of art. Understanding the location, place, and identity of a work of art can help to provide a more nuanced and complete understanding of its meaning and significance. Moreover, when scientific research methodologies are combined with those of the digital humanities, new insights into global cultural heritage may be produced through problematizing the accepted canon and terminology still in use (Table 1.2). The framework thus presented explores the intersection of digital embodiment, digital visual culture, and digital cultural heritage

Table 1.2 Location-presence, place-immersion, and identity-interactivity

Factor/ concept	Referent	Cultural heritage insight
Location	Geographical location where work was created/currently is located	Role played in geographical location to the feeling of presence to reveal intended or original context of work
Presence	Ability of user to feel physically present in a virtual environment	
Place	Symbolic or cultural significance of a location	Ability of works/monuments to transcend the literal and embody cultural hegemony and collective belief systems
Immersion	State of distinction loss between the real and virtual worlds	
Identity	Individual or collective identity of the artist or viewer represented/constructed in a work	Reveal way in which identity, both individual and collective, is enacted through digital embodiment
Interactivity	Degree to which the user can interact with/influence the virtual environment	

through XR technology, specifically VR. By aligning the XR factors of Immersion, Presence, and Interactivity with the art historical concepts of *Location*, *Place*, and *Identity*, a new understanding of art and culture in a digital context emerges.

Through a theoretical exploration of each factor, we delve into how immersion leads to a deeper understanding of historical location, as users are transported to virtual environments that recreate past spaces and places. Presence reveals cultural-historical place by allowing users to feel physically present in these virtual locations, fostering a sense of connection and engagement with the represented cultural heritage. Interactivity enables users to actively engage with and influence the virtual environment, helping to enact identity and personal agency within the digital realm. The treatment builds upon canonical scholarship in the field, offering readers a comprehensive understanding of how XR technology transforms our understanding of art, culture, and history. By examining the ways in which immersion, presence, and interactivity intersect with location, place, and identity, new insights and epistemological models for the digital humanities are uncovered. Furthermore, recent advancements in technology, such as artificial intelligence, machine learning, computer vision, and natural language processing, have made the creation of immersive digital experiences more accessible and efficient. The laborious processes of 3D modeling and photogrammetry have been streamlined, enabling the rapid generation of virtual worlds and cultural artifacts. Additionally, the integration of spatial computing enhances these experiences, allowing users to engage with digital environments in a more intuitive and immersive manner.

In the following sections, each theme shall be explored in depth through case studies and examples of current and future use cases to illustrate the transformative potential of XR technology in the field of art and cultural heritage. By embracing these new ways of seeing and experiencing, we can broaden our understanding of art, culture, and history, gaining a fresh perspective on our own place within the ever-evolving digital metaverse.

References

Amselle JL, Mellott N, Van Dam J (2003) Primitivism and postcolonialism in the arts. MLN 118(4):974–988

Ball M (2022) The metaverse: and how it will revolutionize everything. Liveright Publishing

Barocchi P (1971) Scritti d'arte del Cinquecento. Riccardo Ricciardi, Milan

Baxandall M (1985) Patterns of intention: on the historical explanation of pictures. Yale University Press

Bender SM, Broderick M (2021) Virtual realities: case studies in immersion and phenomenology. Springer Nature

Bhatt M, Freksa C (2015) Spatial computing for design—an artificial intelligence perspective. In: Studying visual and spatial reasoning for design creativity. Springer Netherlands, pp 109–127

Bose N (2009) Sharuk and Shylock: the creation of a South Asian American aesthetic. South Asian Popul Cult 7(3):195–209

Champion E, Foka A (2020) Art history, heritage games, and virtual reality. In: The Routledge companion to digital humanities and art history. Routledge, pp 238–253

Clini P, Ruggeri L, Angeloni R, Sasso M (2018) Interactive immersive virtual museum: digital documentation for virtual interaction. Int Arch Photogramm Remote Sens Spat Inf Sci 42:251–257

Dear M, Ketchum J, Luria S, Richardson D (eds) (2011) Geohumanities: art, history, text at the edge of place. Routledge

Derrida J (1987) Restitutions of the truth in pointing. University of Chicago Press

Easterling PE (2006) Sophocles: the first thousand years. Bull Inst Class Stud:1–15

Favro D (2006) In the eyes of the beholder: virtual reality re-creations and academia, pp 321–334

Gibbs JK, Gillies M, Pan X (2022) A comparison of the effects of haptic and visual feedback on presence in virtual reality. Int J Hum Comput Stud 157:102717

Gruber S (2019) Cultural heritage offences: a view from Asia. In: The Palgrave handbook on art crime, pp 503–525

Hager NN (2020) From nations to networks: an analysis of North American world art historical discourse during the transcultural turn, and the rise of the museum as agent. Doctoral dissertation, University of British Columbia

Han B, Pathak P, Chen S, Yu LF (2022) CoMIC: a collaborative mobile immersive computing infrastructure for conducting multi-user XR research. IEEE Netw 37(6):124–131

Harris J (2002) The new art history: a critical introduction. Routledge

Hegel GWF (2018) Hegel: the phenomenology of spirit. Oxford University Press

Heidegger M (1971) On the way to language

Hutson J, Olsen T (2022) Virtual reality and learning: a case study of experiential pedagogy in art history. J Intell Learn Syst Appl 14

Joyeux-Prunel B (2019) Art history and the global: deconstructing the latest canonical narrative. J Glob Hist 14(3):413–435

Kohl AT (2012) Revisioning art history: how a century of change in imaging technologies helped to shape a discipline. Visual Resour Assoc Bull 39(1)

Leung WK, Cheung ML, Chang MK, Shi S, Tse SY, Yusrini L (2022) The role of virtual reality interactivity in building tourists' memorable experiences and post-adoption intentions in the COVID-19 era. J Hosp Tour Technol 13(3):481–499

Luther W, Baloian N, Biella D, Sacher D (2023) Digital twins and enabling technologies in museums and cultural heritage: an overview. Sensors 23(3):1583
Maurice ML Jr (2022) On the decolonial beginnings of Edward Said. Mod Intellect Hist 19(2):600–624
Melo M, Coelho H, Gonçalves G, Losada N, Jorge F, Teixeira MS, Bessa M (2022) Immersive multisensory virtual reality technologies for virtual tourism: a study of the user's sense of presence, satisfaction, emotions, and attitudes. Multimed Syst 28(3):1027–1037
Melville S (1990) The temptation of new perspectives. October 52:3–15
Milz M (2019) Revisiting Huxley's dystopic vision of future cinema, the feelies: immersive experiences through contemporary multisensory media. AM Časopis Za Stud Umetnosti Med 20:27–42
Newton HT, Spencer JR (1982) On the location of Leonardo's battle of Anghiari. Art Bull 64(1):45–52
O'Connell C, McCauley J, Herbert L (2022) Improvisation-based workshop to build empathy in mentor-mentee relationships and support academic equity. J Stud Aff Res Pract 59(1):87–100
Pope H (2018) Virtual and augmented reality in libraries. ALA TechSource
Pownall M, Harris R, Blundell-Birtill P (2022) Embedding global citizenship in the undergraduate curriculum: a case study from psychology. In: A research agenda for global higher education. Edward Elgar Publishing, pp 211–226
Quang TD (2022) Is living heritage ignored? Revisiting heritage conservation at Cham living-heritage sites in Vietnam. Herit Soc 15(1):46–74
Riegl A (1985) Late Roman art industry. G. Bretschneider
Roussou M (2001) Immersive interactive virtual reality in the museum. In: Proceedings of TiLE (trends in leisure entertainment)
Ruiz-Ruiz IN, Teruel-Serrano MD, Miranda-Sánchez SI (2022) Visual identity based on ancestral iconography: a strategy for re-evaluation of the Caranqui cultural heritage in the Gualimán archaeological site (Ecuador). Heritage 5(4):3463–3478
Schapiro M (1968) The still life as a personal object—a note on Heidegger and van Gogh. In: The reach of mind: essays in memory of Kurt Goldstein. Springer Berlin Heidelberg, Berlin, Heidelberg, pp 203–209
Soussloff CM (ed) (1999) Jewish identity in modern art history. University of California Press
Tally RT Jr (2022) Snow crash. In: Dystopian states of America: apocalyptic visions and warnings in literature and film, p 281
Tzortzaki D (2001) Museums and virtual reality: using the CAVE to simulate the past. Digit Creat 12(4):247–251
Valaskova K, Machova V, Lewis E (2022) Virtual marketplace dynamics data, spatial analytics, and customer engagement tools in a real-time interoperable decentralized metaverse. Linguis Philos Invest 21:105–120
Winckelmann JJ (1987) Reflections on the imitation of Greek works in painting and sculpture. Open Court, La Salle, IL
Wölfflin H (2009) Principles of art history. Göteborgs Universitet
Zhao J, Guo L, Li Y (2022) Application of digital twin combined with artificial intelligence and 5G technology in the art design of digital museums. Wirel Commun Mob Comput 2022

Chapter 2
Macrocosm of History: Everywhere. Everything. Everyone

Abstract The chapter examines how virtual and augmented reality technologies facilitate a profound re-engagement with historical narratives, transcending traditional limits of time and space to offer a more holistic view of the past. The chapter highlights the use of these technologies in creating interactive, immersive experiences that bring historical events, figures, and contexts to life, allowing for a multifaceted exploration of history from multiple perspectives. This chapter also discusses the concept of the "infinitely expanding shape of history," which suggests that history is not static but continuously grows with new discoveries and interpretations. It emphasizes the role of digital technologies in spatializing knowledge, thus allowing for a macrocosm of interconnected historical narratives that reflect complex global interrelations and dependencies. Through virtual museums and interactive digital platforms, users can now traverse vast historical landscapes, experiencing the confluence of art, culture, and technology in shaping historical consciousness. The chapter proposes that these digital experiences can democratize history, making it accessible to a broader audience and ensuring a more inclusive representation of human civilization's diverse heritage.

2.1 The Infinitely Expanding Shape of History

The study of history has long been an essential endeavor, capturing the collective memory of humanity and offering insights into the complexities of our shared heritage. From the ancient Greek historian Herodotus to contemporary scholars, historians have toiled tirelessly to piece together the fragments of time and construct narratives that illuminate the human condition and experience. The task of writing history is a formidable one that includes the challenges of interpreting and contextualizing historical events, movements, and cultural expressions. The task can also never be completed due to the continuous march of time, new discoveries, and changing historical, critical, and methodological perspectives and agendas (White 2005).

J. Hutson, *Art and Culture in the Multiverse of Metaverses*, Springer Series on Cultural Computing, https://doi.org/10.1007/978-3-031-66320-8_2

In practice, history is the systematic study and interpretation of past events, ideas, and experiences that have shaped human societies. It encompasses a multifaceted approach involving the meticulous collection, analysis, and interpretation of diverse sources of evidence, including written documents, material artifacts, oral traditions, and archaeological discoveries. By engaging with these sources, historians strive to uncover the underlying causes, contextualize the occurrences, and discern the far-reaching consequences of historical phenomena. In doing so, they shed light on the intricate development and evolution of cultures, civilizations, institutions, ideologies, and individuals across the expanses of time. Through a critical examination of both primary and secondary sources, historians meticulously reconstruct narratives and offer insightful interpretations that contribute to our collective comprehension of the human past and its enduring relevance in the present (Lowery 2015).

Each historical account, however, is a product of selective choices made by the historian. The inclusion of one historical figure, event, or location inevitably necessitates the exclusion of others. Early modern and modern history, having emerged within the framework of glorifying European patriarchal civilization, traditionally presented itself as a heroic narrative (Jones 2020). Even in our present efforts to revise these historical narratives to encompass the achievements, culture, and contributions of non-Western societies, the prevailing form of history remains that of a story—a method of communication and persuasion that has withstood the test of time. Within any story, certain elements must be present: a beginning, middle, and end; a setting populated by characters, some portrayed as protagonists while others assume antagonistic roles; context, character development, conflict, and resolution, all leading to a discernible moral or ethical takeaway (Byrskog 2022). The manner in which the story is crafted depends on the historian's access to information, documents, artifacts, and sources at a specific time and place. Historically, this has resulted in narratives characterized by narrow plotlines and a limited ensemble of characters, unfolding in a linear and chronological fashion. In reality, however, the tapestry of human existence is infinitely complex, comprising a diverse array of interconnected civilizations, each with its own multifaceted story waiting to be told.

George Kubler's seminal work, *The Shape of Time: Remarks on the History of Things* (1962), serves as a pivotal milestone on the path to the realization of the complexity of history (Kubler 2008). Through his reconsideration of how historians represent time, Kubler challenges the conventional notion of historical change, urging us to transcend the confines of narrow stylistic analysis and embrace a more expansive perspective that encompasses the entire spectrum of cultural production. According to Kubler, historical transformation transcends isolated moments of stylistic shifts; rather, it emerges as a result of interconnected manifestations within a broader temporal framework. By embracing this comprehensive view, historians can transcend the limitations of traditional approaches and delve into the underlying processes and contexts that shape the unfolding of history. Kubler's call to broaden our understanding of historical change presents an insightful framework that illuminates the dynamic interplay between cultural production, societal forces, and the evolution of human civilizations.

Accordingly, modern scholarly discourse embraces an inherent pluralistic approach to historical phenomena, seeking to address the diverse array of contexts that shape the human experience (Rodríguez-Ortega 2019). The abundance of information available to researchers in the contemporary era, spanning primary and secondary sources, big data, and the computational power of artificial intelligence, presents both immense opportunities and daunting challenges. Unsurprisingly, historical scholarship predating the postmodern Information Age predominantly adopted a diachronic perspective. These grand narratives, such as Will Durant's monumental eleven-volume work *The Story of Civilization* (1931), ambitiously aimed to encapsulate the entirety of human history, albeit with varying degrees of access to primary sources and a predominantly Western perspective (Tumadi et al. 2021). However, as scholars began to interrogate these traditional methodologies, there emerged a shift toward more focused synchronic investigations, often limited to the examination of specific historical figures, events, or phenomena, such as Deborah Valenze's *The Invention of Scarcity: Malthus and the Margins of History* (2023). Nevertheless, even within the confines of such narrow inquiries, historians must grapple with the inherent limitations of their research and the epistemological challenges of knowing itself.

2.2 The Microcosm-Macrocosm Pair: Spatializing Knowledge

The expansive breadth of accumulated human knowledge can be truly overwhelming. While it is remarkable to consider that 90% of all data in history has been generated within the past two years, it is worth acknowledging that the ancients recognized the need to devise systems that could aid in the classification of the vast universe (Gupta 2023). Their early attempts at classification would serve as the foundation for further developments by Renaissance thinkers, who sought to refine and expand upon the interconnected elements found in both nature and the cosmos. Through their efforts, a comprehensive epistemological model emerged, capable of providing explanations for a wide array of terrestrial and celestial phenomena, ultimately becoming a ubiquitous framework for understanding the intricacies of the natural world.

The system developed by the ancient Greeks and later adopted by the Romans was rooted in the principles of mathematics, beauty, and harmony. The belief that beauty was intricately linked to harmonious arrangements found its origins in the ancient treatises on architecture and music. Specifically, figures such as Vitruvius and the Pythagoreans emphasized the quantifiable value of harmonious arrangements, guided by mathematical relationships, in achieving beauty (Tatarkiewicz 1972). These early Greek formulations posited that beauty could be expressed through the proportional relationships, arrangement, size, equality, and number of the parts, as well as their interrelationships. One of the most notable and detailed expositions on this valuation comes from Vitruvius, the Roman architect of the first century B.C.E. In his influential work, *De architectura*, Vitruvius argued that the beauty of a building is realized when all its constituent parts adhere to the appropriate proportions. Moreover, the analogy between a building and the human body played a crucial role in understanding the principles of beauty. Vitruvius contended that the human body, like a well-designed building, exhibited a harmonious balance. He famously stated that "the skull, from the chin to the upper brow and hairline, makes up one-tenth of the entire length of the body," underscoring the significance of proportion in both architectural and human form (Tatarkiewicz 1972, p. 169).

As Vitruvius expounded further, he introduced the proportions for a well-formed human figure that would later serve as inspiration for Leonardo da Vinci (1452–1519) and his *Vitruvian Man*, seen in Fig. 2.1 (Morgan and Warren 1914). However, the significance of Vitruvius' formulation extended beyond the pursuit of beauty through the human anatomy; it also offered a means to comprehend the universe itself. Vitruvius observed that a perfect circle could be inscribed using the navel of a figure lying on his back with extended arms and legs as the central point, where the fingertips and toes touch the circumference. He articulated this notion by stating, "And just as the human body yields a circular outline, so too a square figure may be found from it. For if we measure the distance from the soles of the feet to the top of the head, and then apply that measure to the outstretched arms, the breadth will be found to be the same as the height, as in the case of plane surfaces which are perfectly square" (Morgan and Warren 1914, p. 73).

The Pythagoreans and Plato regarded the circle and the square as perfect forms, considering them to be reflections of the cosmos. Thus, in this context, the human being is understood as the microcosm, mirroring the universe, or macrocosm. The proper canon of proportions applied to human bodies, expressed in numerical terms, not only yields mathematical harmony but also reveals insights into the shape of the cosmos itself. Through the interplay of geometric forms and human anatomy, the inherent harmony and interconnection of the microcosm and macrocosm are unveiled, offering a profound understanding of the relationship between human beings and the broader universe.

The conceptualization of spatializing knowledge and mapping the cosmos is also not limited to Western tradition. For example, *mandalas*, which are geometric configurations of circles and squares, serve as symbolic representations of the ideal universe and are considered devotional images within both the Hindu and Buddhist traditions. Beginning in the fourth century, these symbols of the universe hold significant spiritual importance and are employed in different spiritual traditions for various purposes, such as focusing attention, establishing sacred spaces, and aiding in meditation and trance induction (O'Neal 2020). In Hinduism, mandalas, also known as *yantras*, take the form of a square with four gates enclosing a central circle. These geometrical diagrams exhibit radial balance and are considered to represent the dwelling place of deities (Fig. 2.2). While being smaller versions of mandalas, yantras are used in sadhanas, puja, and meditative rituals to invoke the presence of deities. In service of spiritual experiences, these designs are believed to reveal cosmic truths (Bühnemann 2003).

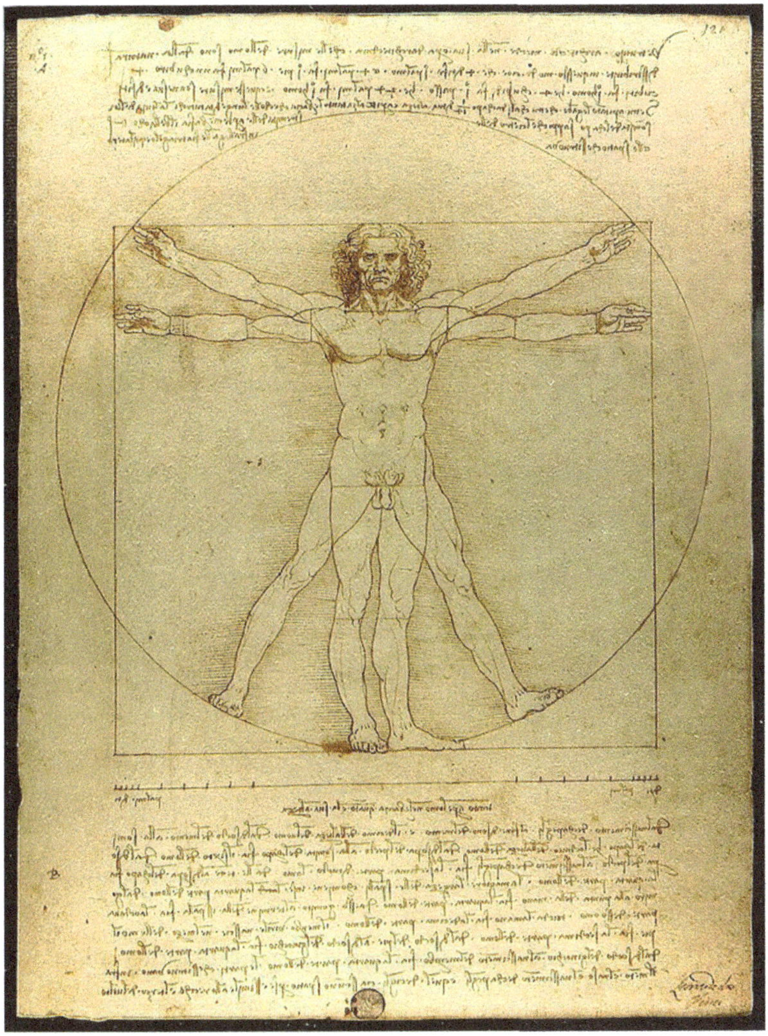

Fig. 2.1 Leonardo da Vinci, Vitruvian Man, ca. 1487 Gallerie dell'Accademia, Venice. Public (from Wikimedia Commons, licensed under CC0)

The similar concept of mandalas extends to Buddhism as well, where they are used in sandpainting and meditation practices, particularly in Vajrayana Buddhism. Mandalas in this context visually represent the core essence of Vajrayana teachings and the nature of the Pure Land or enlightened mind; they can depict various aspects of enlightenment and are associated with specific Buddhas or wisdom forms (Geravand and Mirzaei 2020). For instance, the mandala of the "Five Buddhas" embodies different aspects of enlightenment and serves as a visual representation of the Mandala of the Two Realms when paired with the mandala depicting the Five

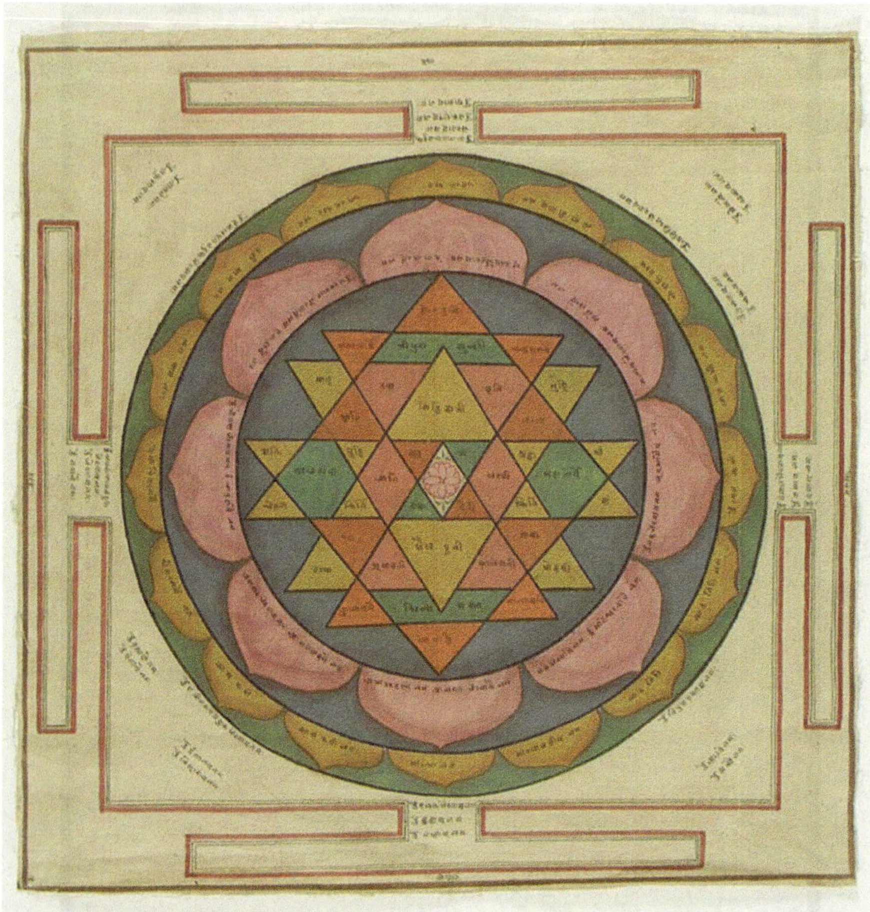

Fig. 2.2 Shri Yantra with explanatory labels applied, nineteenth century (from Wikimedia Commons, licensed under CC0)

Wisdom Kings (Ma 2020). Buddhist mandalas also symbolize wisdom and impermanence. The outer circle often represents wisdom, while the ring of eight charnel grounds serves as a reminder of the transient nature of life and the impermanence of *samsara*, represented in Fig. 2.3, the continuous cycle of life, death, and rebirth. The mandala palace, depicted within the rings, is populated by deities and Buddhas.

Mandalas can also depict the entire universe, with Mount Meru—the sacred five-peaked mountain—as the central *axis mundi*, surrounded by continents. The interconnectedness between the outer world and inner human consciousness is represented through the ambivalent resonance of the symbols, but Buddhist mandalas are not merely visual images: they are worlds to be explored and experienced. These cosmic maps follow a quincunx pattern, with a central point surrounded by four directions, each presided over by a color-coded Buddha representing a psychological

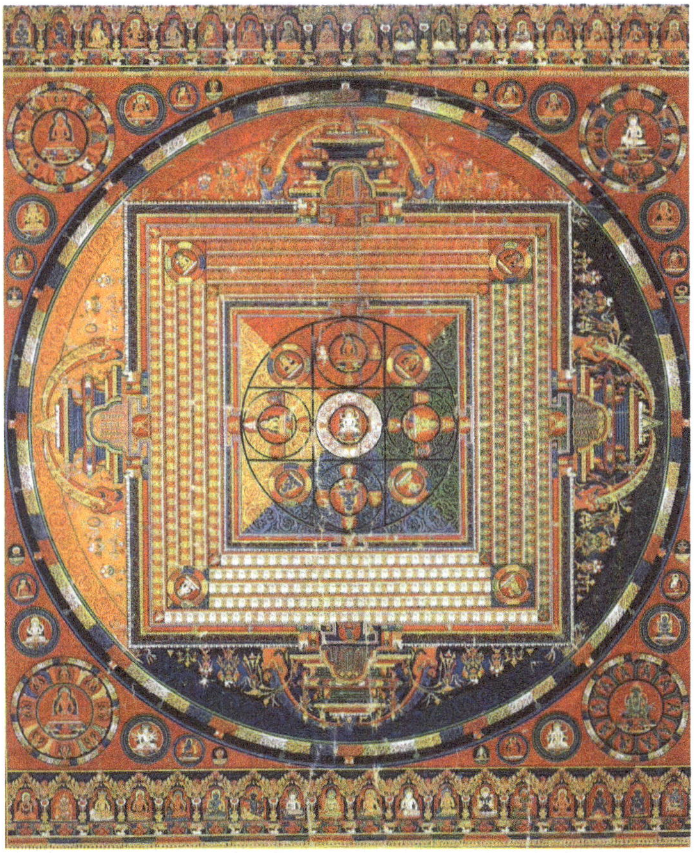

Fig. 2.3 Anonymous Tibetan Artist, Mandala of Vajradhatu, thangka painting, nineteenth century (from Wikimedia Commons, licensed under CC0)

defect and its corresponding insight. By reciting mantras and forming mudras (ritual gestures) associated with these Buddhas, practitioners transform their defects, which allows them to approach enlightenment (Winfield 2019). Thus, the circle-square/microcosm-macrocosm pairing can be found in disparate cultures and demonstrates the tendency of humans to construct conceits to organize not only their relationship to the cosmos and the divine, but also to provide a mnemonic machine to facilitate that understanding. In the West, the seemingly straightforward analogy of man as a microcosm, revealing a map of the macrocosm, belies the intricate systems encapsulated within the circle-square design. This epistemological model, in fact, comprises a complex network of correspondences that evolved to categorize the interconnections between human beings and the natural world. Originating from the ancient Greeks, who held a deep reverence for the number four, all phenomena were organized into a series of quadripartite schemas, each with overlapping relationships that allowed for superimposition.

At the macrocosmic level, the universe was believed to consist of four fundamental elements, which also constituted the human body. Through the transformative process of digestion, these elements were converted into the four humors within the organs. Furthermore, these elements were associated with fluids and their corresponding qualities: hot/warm, dry, cold, and wet/moist. The celestial bodies, including the stars, planets, and seasons, were similarly influenced by these relationships, as the changing seasons mirrored the stages of human life. Thus, the humors corresponded to the four elements of the earthly realm, which, in turn, corresponded to the four ages of man, as well as the twelve zodiac signs divided into four groups of three, as seen in Fig. 2.4. This intricate web of interrelationships extended to encompass all aspects of the universe and the nature of humanity, highlighting the interconnectedness of all things. Given the complexity of these interrelationships, authors of the time developed mnemonic devices to assist their readers in comprehending these overlapping classifications. These devices often served as structuring conceits, offering a framework to explain the intricate interplay of correspondences to a wider audience. The utilization of mnemonic devices was essential in conveying this elaborate system of interconnections, ensuring that the complexities of these overlapping classifications could be grasped and understood by a diverse range of individuals.

One such mnemonic machine would arrive in 1584, when Giovan Paolo Lomazzo (1538–1592) published his massive encyclopedic study on painting, *Trattato dell'arte della pittura, scultura ed architetta* (*Treatise on the Art of Painting, Sculpture, and Architecture*). This exhaustive work consisted of seven chapters, each dedicated to one of the seven fundamental aspects of painting. To elucidate the intricate correspondences within these chapters, the Milanese theorist also released his *Idea del tempio della pittura* (*Idea of the Temple of Painting*) in 1590. In this treatise, Lomazzo transformed the divisions of the Trattato into a metaphorical "Temple of Painting," employing architectural elements as pillars to symbolize the planetary temperaments and their respective governors, including renowned artists such as Michelangelo, Leonardo, and Andrea Mantegna. Lomazzo openly acknowledged that the design of this temple was inspired by Giulio Camillo's (ca. 1480–1544) *L'Idea del theatro*, which was published posthumously in 1550. The theatre of Camillo aimed to encapsulate the entirety of the universe, employing a Vitruvian plan that featured a large auditorium divided by seven aisles, each dedicated to a planetary deity (Yates 1966). These aisles were further divided into seven rows, ascending from the floor level to the upper tiers (much like those from Vitruvius' design seen in Fig. 2.5), representing the terrestrial and celestial spheres of the world. Each row and aisle featured doors adorned with significant depictions that triggered memories of specific locations within the universe.

The connection of the heavens and earth in architectural form also finds resonance in Asian culture. As with Christian churches, that highlight the symbolism of the cross through the crossing of the nave and transept, so too are geometrical designs used to provide physical access and navigate the universe. For example, both Buddhist and Hindu temples continue the symbolism found in geometric mandala, though to different ends. Upon entering a Hindu temple, such as the Kandariya Mahadeva Temple in Khajuraho (Madhya Pradesh, India) (ca. 1030), seen in Fig. 2.6, individuals

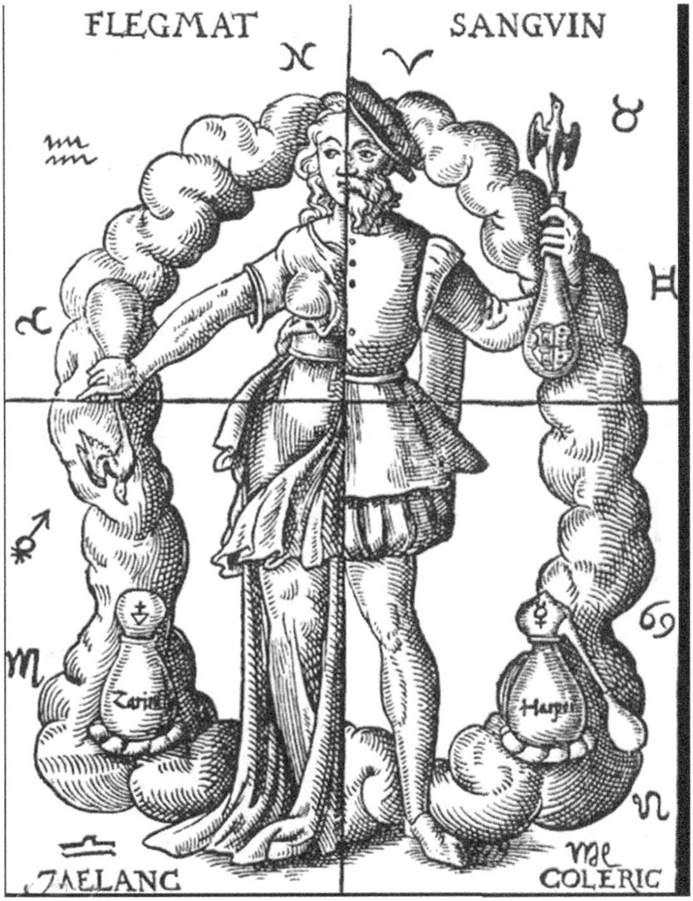

Fig. 2.4 Leonhard Thurneysser, Quinta Essentia, woodcut, 1574 (from Wikimedia Commons, licensed under CC0)

are transported to a realm that mirrors the intricate organization of the cosmos. These northern temples, reminiscent of the mountainous terrain where they are located, signify the connection between the earthly and divine realms (Das and Patnayaka 2020). Elaborate sculptures of deities embellish the entrance gateways leading to a meticulously designed interior space, crafted according to sacred geometric principles. The layout follows a mandala-like pattern, characterized by concentric squares and circles that create a serene and balanced atmosphere, while at the heart of the temple lies the central shrine, known as the *garbhagriha* or *garbha griha*. This "womb chamber" symbolizes the axis mundi, the cosmic center, where the divine energy resides (Kumar 2021). The temple, in its entirety, becomes a microcosm—a sacred narrative that mirrors the grandeur and meticulous organization of the cosmos.

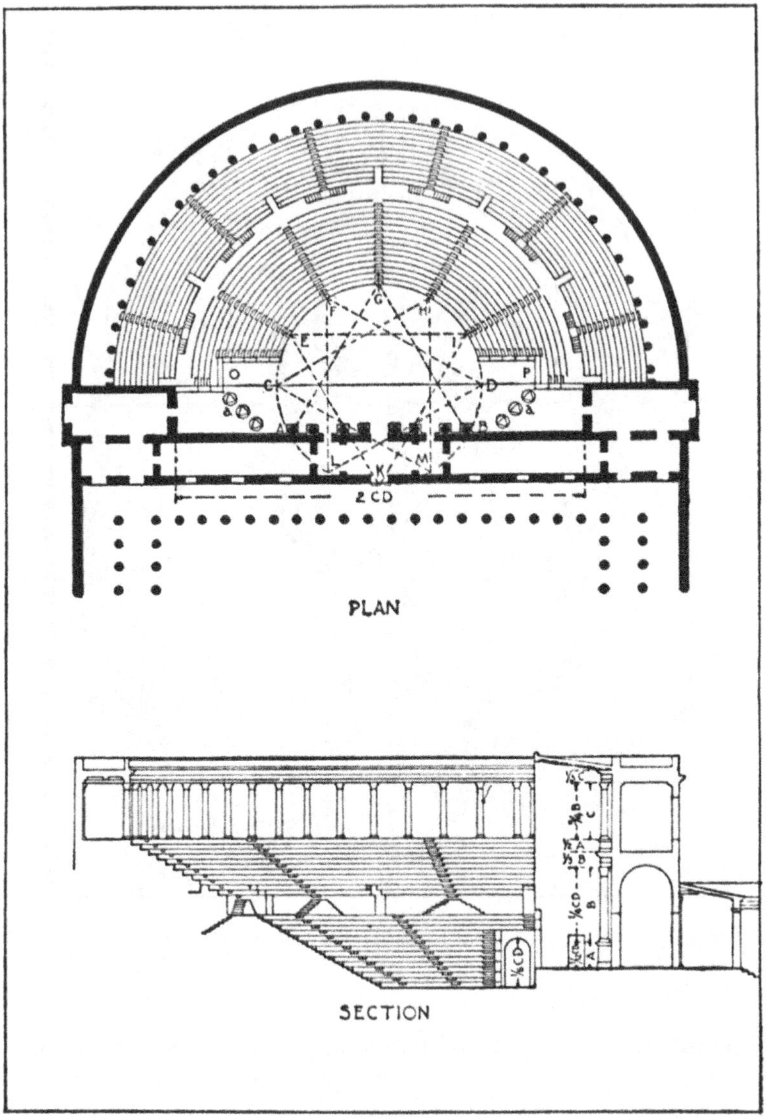

Fig. 2.5 Roman Theatre, Vitruvius, The Ten Books on Architecture. Morris Hicky Morgan trans (from Wikimedia Commons, licensed under CC0)

In addition to Hindu temples, Buddhist mandalas are also manifested in architectural forms, exemplified by the Borobudur temple in Indonesia, seen in Fig. 2.7. Constructed during the ninth century under the Shailendra dynasty, Borobudur is an immense Buddhist monument situated in Central Java, renowned as one of the most magnificent Buddhist structures (Kowal 2019). Reflecting the essence of Buddhist

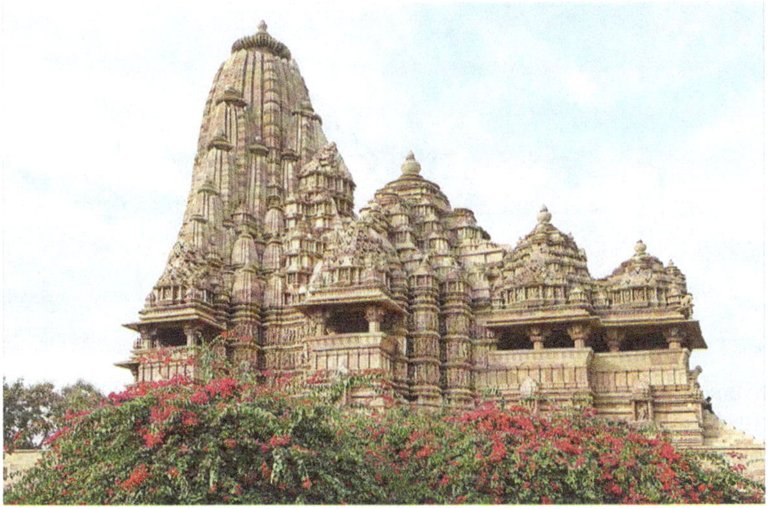

Fig. 2.6 Vidyadhara, The Kandariya Mahadeva Temple, ca. 1030, Sevagram, Khajuraho, Madhya Pradesh, India (from Wikimedia Commons, licensed under CC-BY 2.0)

cosmology, the temple is designed as a colossal mandala comprising nine stacked platforms, with the central stupa adorning the highest tier. In fact, the entire architectural layout mirrors the mandala, featuring a square base that symbolizes the earthly realm. Elaborate relief carvings grace each platform, depicting narratives from Buddhist teachings and cosmological imagery. Progressing through the levels, the intricate reliefs serve as spiritual guides, illustrating the journey to enlightenment. The outer layers portray vibrant scenes from daily life, offering glimpses into the human experience. As visitors ascend, the reliefs transition to represent Buddhist philosophies, directing individuals symbolically toward the path of enlightenment. Reaching the summit marks the culmination of this symbolic pilgrimage, where the central *stupa* dominates the vista, embodying the transcendent realm and the ultimate goal of Buddhist practice (Zaluchu and Widjaja 2019).

These religious structures and the theatre metapor, therefore, as a whole were meticulously designed to enable the human mind to comprehend nothing less than the macrocosm. The theatre, in particular, served as a "mnemonic tool," originally intended to provide orators with well-organized material for their thoughts. The concept comprised a system of "places and images" precisely arranged, "sufficient for the location and management of all human concepts and of all things in this world, not only those which belong to the various fields of knowledge, but also to the noble arts and the mechanical" (Bernheimer 1956, p. 229). By carefully classifying all knowledge and establishing an ideal order for all possible topics, the human mind itself could become a perfect reflection of the macrocosm. The neoplatonic belief in cosmic influence, with its notion of a harmonious correspondence among various links in a chain, deeply influenced Lomazzo's (1973) conception of the Temple of Painting. Just as the ascending structure of Camillo's theatre and that of

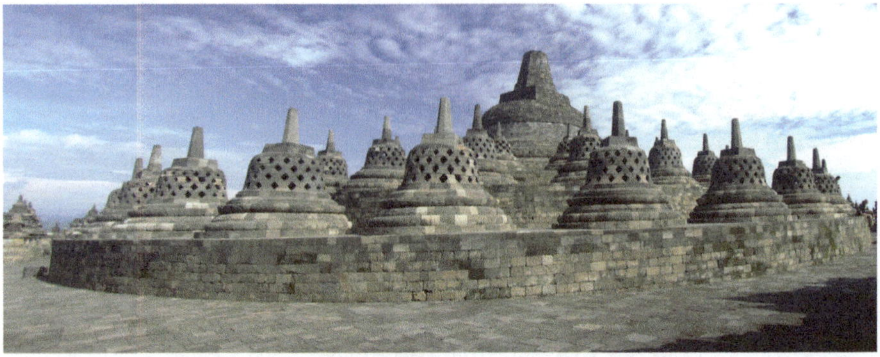

Fig. 2.7 Borobudur Temple, ninth century, Central Java, Indonesia (from Wikimedia Commons, licensed under CC-BY 2.0)

Borobudur symbolized the journey from the earthly to the heavenly realms, Lomazzo also embraced the idea of cosmic interconnectedness, acknowledging the harmonious relationships that permeated different aspects of the artistic realm.

The organizing principles at play in the Theatre and Temple would influence many early modern authors. One influential example was published by Giovanni Paolo Gallucci (1538–1621), the preeminent astronomer throughout Europe, thanks to his *Theatrum mundi et temporis* published in Venice in 1588 (reprinted in 1589) (Bohern 1969). This work emerged during a period of heightened interest in celestial mapping in the latter half of the sixteenth century, driven in part by the initiatives of popes such as Pius IV and Gregory XIII (Hess 1967). The publication arrived shortly after the issuance of the papal bull *Coeli et terrae* by Sixtus V in 1586, which prohibited the practice of astrology and other forms of divination. In response, Gallucci reaffirmed the significance and nobility of celestial science, emphasizing the importance of understanding it free from any elements of superstition. He proceeded to discuss the nature and qualities of each planet, their radiation, and their influence on human bodies through their positions in the zodiac.

These celestial influences, according to Gallucci, could be discerned by examining the distinguishing characteristics of the planets, such as their masculine or feminine nature, their commanding or obedient disposition, and their potential for hostility (Thorndike 1923). Notably, Gallucci highlighted that the effects of particular zodiac signs extended beyond general influences on human bodies based on proximity, which had been well-circulated in the previous century, and already well-established in the arts as seen in the Limbourg Brothers illustration in Fig. 2.8. He enumerated how individual members of the human body could also be impacted. As a result, astrological diagrams or figures of constellations were drawn up, serving as crucial tools in medical practice to determine the appropriate timing and type of treatment to administer. Furthermore, the zodiac sign under which a patient was born held paramount importance, as it would determine pre-existing conditions and the dominant influences associated with that sign.

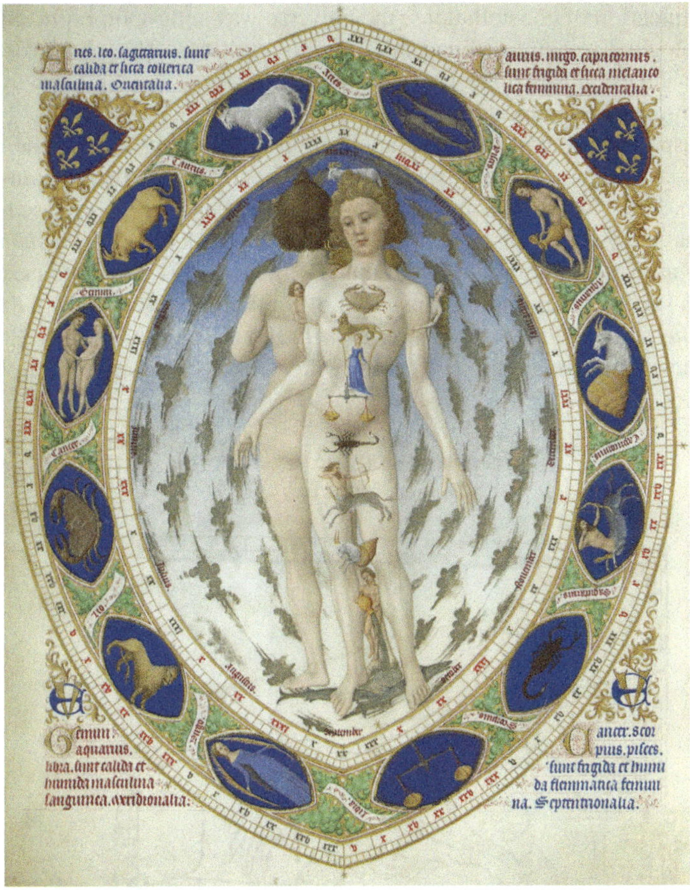

Fig. 2.8 Limbourg Brothers, Anatomical Man, Très Riches Heures du Duc de Berry (1412–1416) (from Wikimedia Commons, licensed under CC0)

In the following century, the work underwent a title change: from *Theatre of the World and Time* to *An Explanation of the Celestial Bodies and the Things Dependent on Them by Means of Instruments, Scrolls, and Figures* (1603, 1605). The updated treatment was divided into six parts, dedicated to describing and mapping the entirety of the celestial and terrestrial realms. Notably, it depicted the inferno as ten concentric circles within the Earth. Serving as the first modern celestial atlas, the maps featured a coordinate and trapezoidal projection system, enabling the accurate determination of star positions derived from the Copernican catalog. Gallucci's intention was to create an informative text accessible to non-specialists, employing simple mathematical demonstrations complemented by appropriate images. Descriptions were presented in the form of pictures, vividly illustrating the subjects under discussion. The work also included forty-eight maps of the Ptolemaic constellations, each accompanied by corresponding mythological figures like Cassiopeia in Fig. 2.9. Through the personification of constellations, the universe was anthropomorphized, echoing the cosmological principles found in Vitruvius' writings. This endeavor aimed to categorize and establish connections between the heavens and the Earth through the microcosm-macrocosm relationship.

Demonstrating the wide-ranging application of these systems in the realms of astronomy, astrology, and the arts, Gallucci also ventured into the domain of artistic instruction. It was within this intersection of subjects and with a diverse readership in mind that the astronomer-translator published his work titled *Della simmetria dei corpi humani* (On the Symmetry of Human Bodies) in 1591. This treatise on human proportion served as an Italian translation of the Latin edition of *Vier Bücher von menschlicher Proportion* (or *Proportionslehre*) (Four Books on Human Proportion) by the renowned German artist Albrecht Dürer (1471–1528) as seen in Fig. 2.10.

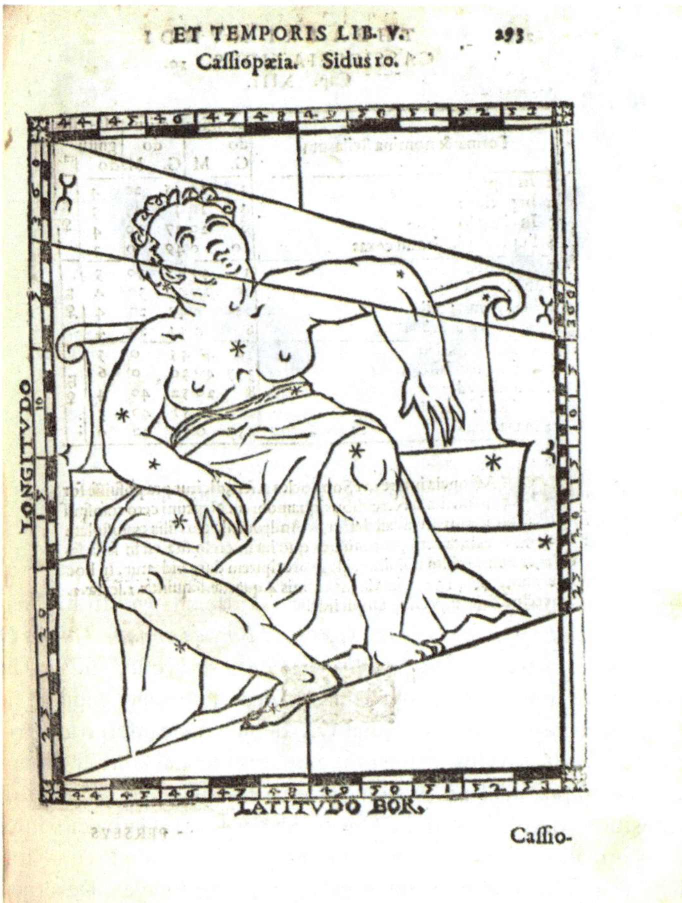

Fig. 2.9 Giovanni Paolo Gallucci, Casiopia, 1588. Woodcut, from Theatrum mundi, et temporis, Venice 1588 (from Wikimedia Commons, licensed under CC0)

Through this translation, Gallucci shed light on the significance of symmetry and proportion within the context of contemporary education, particularly for artists. Right from the beginning of his treatise, Gallucci established the premise that human proportions (the microcosm) are a reflection of the harmony inherent in the universe (the macrocosm). The Vitruvian pairing, which demonstrated the numerical calculation of the universe through the human body, vividly illustrated the perceived interconnectedness between the cosmos and the human form.

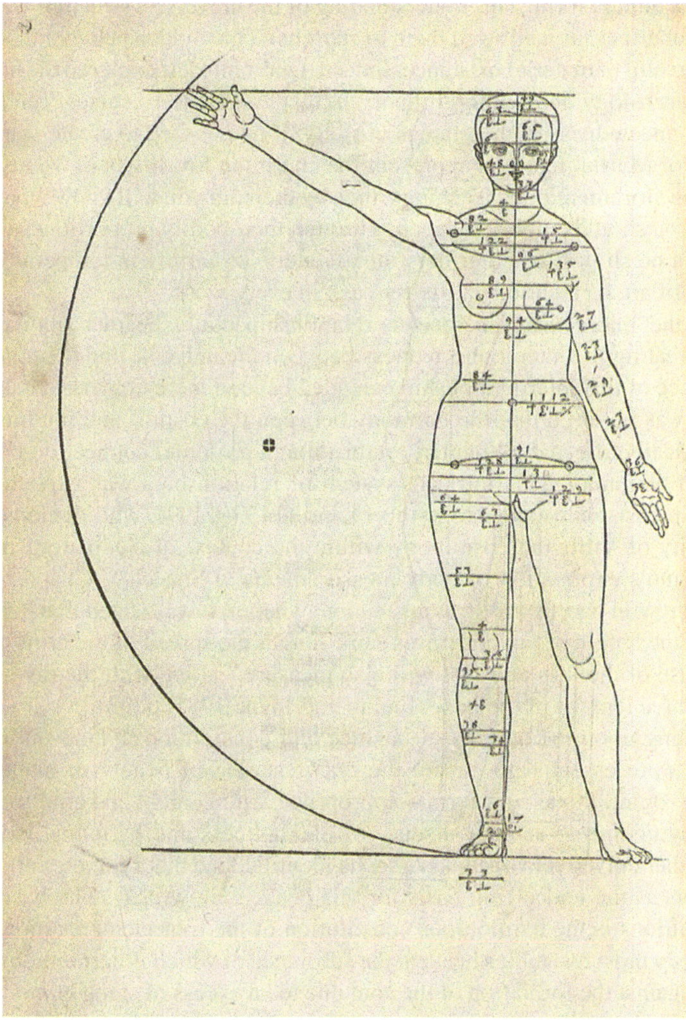

Fig. 2.10 Giovanni Paolo Gallucci (after A. Dürer), Proportions of a Woman, 1591. Woodcut from Della simmetria dei corpi humani, libri quattro, Roberto Meietti, Venice 1594 (from Wikimedia Commons, licensed under CC0)

This cosmological perspective established a profound connection between the balanced composition of human figures and the theory of the four elements and humors. These physiognomic and humoral interpretations found their classifications within the framework of astronomy and astrology, an apt reflection of the translator's background. Moreover, the endeavors to categorize and map the human body, as exemplified by Dürer and other artists, were perceived by art theorists, astrologers, and cosmologists as a means to unveil the macrocosm of the universe, thereby revealing its inherently "divine" and harmonious beauty within the microcosm of the human form. This understanding of the universe's structure was deemed essential for artists, as it allowed them to apprehend the hidden beauty inherent in the imperfect realm of material existence. Indeed, Dürer himself explored the intertwined realms of astrology and humoral theory in his "Aesthetic Excurses," underscoring the interconnectedness of these areas of study. If one wished to create some strange Saturnine or Martial figure or represent the charming loveliness of Venus, and you practice the aforementioned teachings, then by their aid you will know what standard and style you should employ for each. Through their bodily proportions, all families of men can be shown, whether fiery, airy, watery, or earthly in temperament. Thus the power of art, as mentioned, master each in every work.

While the microcosm-macrocosm relationship could be quantitatively represented, revealing the interconnectedness between the universe and the human body, the influence of this interrelationship extended beyond mere numerical calculations. In fact, it was believed that the harmony between the cosmos and the human form sympathetically affected all matter, establishing a profound connection between all things and humanity. This intricate system of relationships was expounded upon by the humanist-philosopher Marsilio Ficino (1433–1499), who demonstrated the applicability of Vitruvian cosmology within the context of Neoplatonism (Hutson 2010). Ficino's exploration of body measurements in Speech V Chapter VI of his *Convivio* drew in part from Vitruvius' canon, wherein it was stated that "eight heads will make the length of the body; this same length the spread of the arms to the side and likewise of the legs and feet will also measure," along with the division of the face into three lengths of the nose (Ficino and Jayne 1985, p. 93).

Expanding upon the notions of balance and constitution, Ficino outlined three steps that were essential to ensure the manifestation of beauty in matter. Within each step—order, mode, and form—appropriate arrangement and equilibrium were deemed paramount to achieve a sense of pleasantness and harmony. However, as Ficino further delved into his discourse, he identified the theory of the four elements and humors as the underlying basis for this beauty. He writes, "The basis of these three conditions is the harmonious constitution of the four elements in such a way that our body most resembles heaven, the substance of which is harmonious and does not rebel against the formation of the soul due to an excess of some humor. In such a way the splendor of heaven easily appears in the body as similar to heaven. And this perfect form of man, which the spirit possesses, will be more integral in peaceful and obedient matter …" (Ficino and Jayne 1985, pp. 93–94).

The relationship between the body and the universe, encompassing the microcosm and the macrocosm, extends beyond geometrical connections and delves into the realm of metaphysical balance. This conception finds its roots in Greek cosmology, where Socrates, in Plato's *Philebus*, articulates the notion that the elements present in the human body are derived from the elements existing in the cosmic body. According to Socrates, "Whence can a human body have received its soul, if the body of the universe does not possess a soul?" (Hahm 1977, p. 138). This belief establishes a parallel between the balanced composition of the world and the universe, consisting of the four elements—earth, air, fire, and water—and the equilibrium of the human body and soul, which is maintained through the four humors—black bile, yellow bile, phlegm, and blood. For physicians influenced by the teachings of Hippocrates and Galen, restoring health involved the reduction of excesses through the introduction of opposing elements, such as bloodletting (Jouanna 2012). In the realms of art, philosophy, astronomy, and astrology, the pursuit of balance and harmony in the universe, human body, and artistic creations was considered essential for achieving order, well-being, and beauty. Thus, according to Ficino, if the necessary preparations for beauty, encompassing these various aspects, are diligently observed, the "splendor of heaven easily appears in the body as similar to heaven" (Ficino and Jayne 1985, p. 94).

2.3 Epistemology of the Cosmos and Universal Mapping

The interconnectedness between humanity and the universe naturally led to the intertwining of astronomy and astrology with the study of human proportion. Scholars, like Gallucci, who were interested in mapping the cosmos, recognized the inherent reflection of the microcosm-macrocosm pairing, whereby knowledge of human beings could be extended to encompass the understanding of the universe itself. This understanding was facilitated by the belief in the dispersion of the four elements throughout the universe, as well as the notion that the universe was structured in a similar manner to human beings. Plato's *Timaeus*, for instance, argued that "The visible universe is a living creature, having soul (*psyche*) in body and reason (*nous*) in soul … Man is also composed of reason, soul, and body; but his body will be dissolved back into the elements, and the two lower parts of his soul are also mortal. Only the divine reason in him is imperishable" (Cornford 1997, p. 38). The visible universe is thus a living entity with a soul in its body and reason in its soul. Similarly, human beings are composed of a body and soul with reason, but only the divine reason within them is imperishable, while the other parts are mortal and destined to return to the elements. This model, appealing to Christian thinkers as well, posited that the human body is formed from the elements and will eventually return to them after death. The constitution of these elements can be observed in the various planets, which, in turn, influence the characteristics of individuals born under their astrological influence. The manner in which celestial bodies exert their influence on earthly entities reflects the organization of the cosmos.

As Elias Ashmole, an antiquarian and founding member of the Royal Society, noted in his work *Theatrum chemicum britannicum* (1652), the universe is divided into supercelestial, celestial, and natural realms, with the sublunary world serving as the final recipient of divine ideas transmitted by the planets and stars. This hierarchical structure reveals the transmission of celestial influences to the earthly realm, illustrating the interconnected relationship between the cosmic order and the world inhabited by human beings (Schuler 1980). The belief in the interconnectedness of the universe and the human body found practical application in treatises on art. In Chapter XXVI of his *Idea del tempio*, Lomazzo drew a connection between proportions and astrology, aligning himself with the Ficinian definition of beauty and the significance of humoral theory. Drawing from Ficino's *De Vita Triplici*, Lomazzo cited the works of mathematicians and astrologers to support his argument as he observed that beauty in the human body is derived from the qualities that make each body unique and distinct from one another. As he writes, "But when it comes to fashioning the bodies, beauty is taken from the qualities that make all our bodies dissimilar, changing from one to the other more or less ..." This diversity is influenced by the elements and planets. According to Lomazzo, "There can be however only four main types of dissimilarity, in accordance with the number of elements and the strength of their qualities, which the mathematicians assert are the bases of all the forms or rather types of human bodies" (Hutson 2010). Mathematicians assert that these bases underlie all the forms or types of human bodies.

Each planet is associated with a specific quality inherent in the elements, which is then manifested in earthly bodies. For example, individuals with Martian types, influenced by the hot and dry nature of fire, tend to have larger limbs. On the other hand, Lunar types, influenced by the cool and humid nature of water, have more restricted limb sizes. Similarly, Jovian types, influenced by the airy quality of air, exhibit full-bodied figures, while Saturnine types, influenced by the earth element, have straight and concave limbs. By linking astrological attributes to bodily proportions, Lomazzo sought to explore the influence of celestial forces on the physical manifestations of human bodies. This understanding further exemplifies the belief in the interconnectedness between the macrocosm and the microcosm, wherein the qualities of the universe are reflected in the diversity of human forms.

Moreover, within the astrological framework, man was regarded as a microcosm, and it was already established that the twelve zodiac signs exerted influence over different parts of the body. Ficino, in his work *De vita triplici*, presented a model of human life wherein thousands of lines of influence connected each organ and aspect of the body and soul to specific stars and planets (Grafton 1999). This interconnection between celestial bodies and human beings was likened to the transmission of light, as Ficino (1989, p. 297) observed: "... every spirit is therefore instantly moved one way or the other and formed by these things. However it turns out, such in turn in some degree it makes the mind, and absolutely such, the quality of the body. Then as soon as it is opportunely exposed to the Graces through things pertaining to them, being also naturally in accord with those things, it immediately gains wondrous gifts of the Graces through their rays, which both flourish everywhere and are akin to it."

In the seventeenth century, Robert Fludd (1574–1637) continued the symbolism of the macrocosm-microcosm relationship in his work *Utriusque cosmi historia* (1617–21). The title page of his work depicts the macrocosm-microcosm, with the right foot positioned in Leo and the left foot in Libra seen in Fig. 2.11. The right hand is placed in Gemini, the left hand in Capricorn, and the head is situated at the Pisces-Aries cusp. Each of these signs is ruled by one of the five classical planets. Hence, the Pisces-Aries cusp corresponds to Jupiter and Mars, Capricorn to Saturn, Gemini to Mercury, and Libra to Venus. The Sun governs Leo, while none of the five planets are associated with Cancer, which is ruled by the Moon. Fludd represents the Sun, Moon, and other planets in the macrocosm, positioned above man in the microcosm. The arrangement of these five principal parts in correspondence with the zodiac spheres led to the adoption of the pentagram as a symbolic representation of man as the microcosm. This arrangement emphasizes the influence of the zodiac spheres on different parts of the body, highlighting the meaningful connections between different areas of the human form and corresponding parts of the universe.

In a similar vein to Fludd's approach, early modern writers employed intricate mnemonic devices to comprehend the vast expanse of the universe and encompass all branches of human knowledge. Works like *Theatrum mundi et temporis* offer valuable insights into the utilization of mnemonic frameworks, especially those used to spatialize knowledge. Through the microcosm-macrocosm pairing, these authors aimed to categorize and establish connections between the celestial and terrestrial realms. Drawing inspiration from other mapping systems, such as Camillo's Theatre, which was discussed earlier, Gallucci's "theatre" and others aspired to encapsulate the entirety of the "world and time." Initially conceived as a "mnemotechnic tool" for orators, this system organized "places and images sufficient for the location and management of all human concepts and of all things in this world, not only those which belong to the various fields of knowledge, but also to the noble arts and the mechanical" (Bernheimer 1956, p. 229). By carefully classifying and ordering knowledge, it was believed that the human mind itself could become a flawless reflection of the macrocosm. Considering the vast and interconnected nature of the diverse fields of knowledge, it is not surprising that mnemonic devices are often employed. Among the most prevalent techniques utilized to facilitate understanding and navigation of this expansive intellectual terrain is the spatialization of knowledge.

Through various cultural and historical contexts, the spatialization of knowledge as a mnemonic device has been a prevalent practice. Not only seen in the architectural traditions of Hindu and Buddhist temples, but also in other domains such as theatre and Western intellectual pursuits. In the realm of theatre, the concept of spatializing knowledge can be observed in the work of influential figures like Sebastiano Serlio (1475–1554), Giulio Camillo, and Giulio Parigi (1571–1635). Serlio, an Italian architect, theorist, and contemporary of Camillo, developed the concept of the "Theatre," an architectural mnemonic device that represented a systematic classification of knowledge. The Theatre was designed as a physical structure divided into compartments, each containing visual representations and symbols associated with

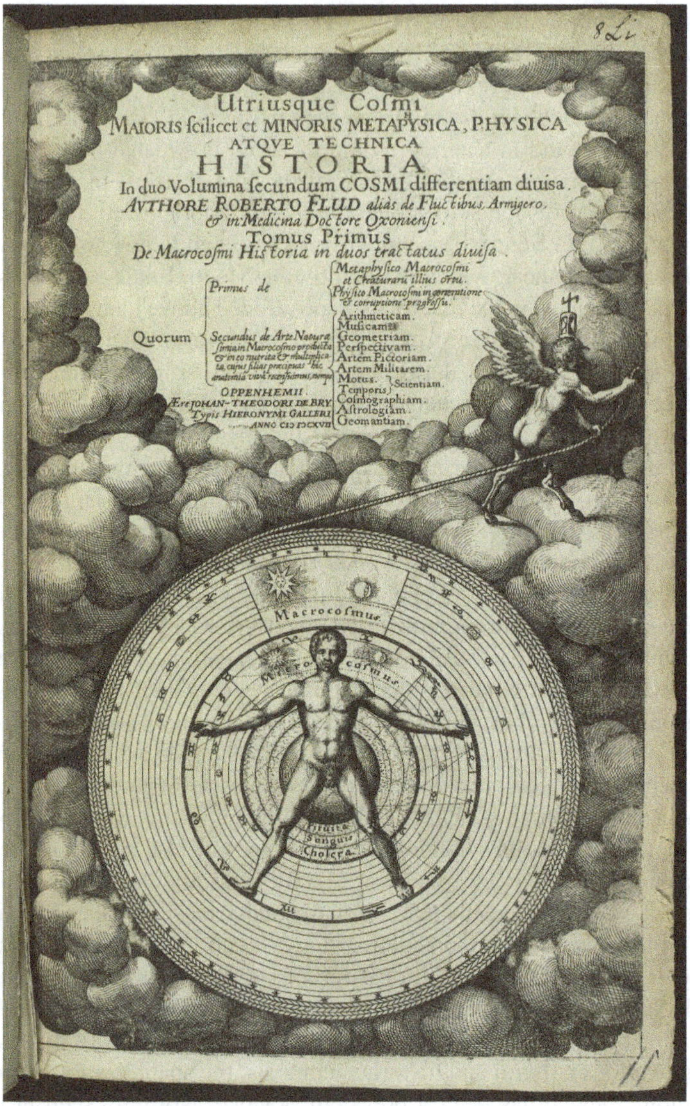

Fig. 2.11 Robert Fludd, title page, 1617. Engraving from Utriusque cosmi historia, Oppenheim, 1617 (from Wikimedia Commons, licensed under CC0)

specific fields of knowledge. By spatially arranging and organizing these compartments, Serlio aimed to facilitate memory and comprehension, providing a tangible framework for the human mind to navigate through complex information.

In the Western intellectual tradition, the inclination to spatialize knowledge as a mnemonic device has also been apparent. The use of mind maps, concept maps, and organizational diagrams are common techniques employed to visually represent and organize information. These spatial representations serve as external aids for the human mind to grasp complex concepts, establish connections, and retain information effectively. From flowcharts to hierarchical structures, the Western penchant for spatializing knowledge reflects the recognition of the human mind's capacity to comprehend and remember information through spatial arrangement. Through the careful classification and ordering of knowledge, whether in the form of temple designs, mnemonic devices like the Theatre by Camillo, or spatial representations in Western intellectual pursuits, cultures throughout history have sought to enhance understanding, facilitate memory recall, and create a harmonious relationship between the human mind and the wider cosmos. The spatialization of knowledge serves as a tangible manifestation of the interconnectedness between the microcosm and the macrocosm, as well as a testament to the enduring quest for comprehensive knowledge organization across diverse cultural contexts.

2.4 The Virtual Museum as Theatre of Memory

The spatial understanding of experience demonstrated through mnemonic mapping and visiting sacred sites to explore the interconnectedness of the cosmos, is now being brought together in the digital realm through immersive technologies. This convergence allows for the attainment of transformative and meditative states akin to those experienced when using mandalas as guides. A notable exhibition at the Asian Art Museum of San Francisco exemplifies this phenomenon by transforming Tibetan Buddhist mandalas into a three-dimensional virtual reality (VR) installation. The project aimed to enhance the visitor experience by creating an immersive and interactive VR installation inspired by the mandalas exhibited. The installation, called *Mandala Flow State*, incorporated elements of the original artwork and utilized neurofeedback to modulate the visual scene. Electroencephalograms (EEGs) were recorded to dynamically adjust the intensity of the fog obscuring the mandala, mimicking the experience of clearing the fog with one's mind during meditation. The collaboration between art and technology demonstrated how scientific tools and concepts can be adapted to create artistic installations that enrich the visitor's experience. The study highlights the potential of VR and neurofeedback in providing contemplative and therapeutic experiences, blurring the boundaries between subject and object, mind and matter.

However, the immersive nature of cultural experiences relies on both space and time, as culture is inherently dependent on specific locations and historical periods. Previously, physical pilgrimage to sacred sites was essential, serving as a tangible expression of devotion or a literal path toward enlightenment, allowing individuals to experience the cosmos firsthand. However, the advent of new technologies has revolutionized our ability to transcend the constraints of time and space. Now, one can effortlessly travel to any corner of the world and explore cultural heritage sites as they exist in the present moment. Yet, this is merely the initial application of accessing cultural heritage and surpassing temporal and spatial boundaries. These innovative technologies enable the recreation of any historical epoch, allowing users to immerse themselves in a human-scale representation complete with visual stimuli, aromas, and tactile sensations. The only limitations lie in the availability of information pertaining to a given location, people, culture, language, geography, and so on. With this expanded potential, a new era of virtual cultural heritage experiences and worlds unfolds, inviting us to delve into the rich tapestry of human civilization throughout the ages.

While VR technologies have been used in applied fields since the 1960s in flight simulators for the military and industry training, it has only been over the past two decades where most progress has been made to utilize the technology for experiencing art and culture (Schroeder 1993). Before the widespread development of virtual learning environments (VLE), such as virtual museums (VM), immersive technologies were initially harnessed to enhance physical museum exhibitions for edutainment purposes. However, the adoption of these technologies by museum curators was limited in the early 2000s, as the understanding and availability of immersive installations were still in their nascent stages (Carrozzino and Bergamasco 2010). Nonetheless, scholars like Roussou (2001) recognized the immense potential of virtual environments for cultural heritage institutions, highlighting the pioneering work of the Foundation of the Hellenic World (FHW) in Athens. FHW utilized CAVE technology, seen in Fig. 1.5, to create digital reconstructions of ancient cities, starting with Miletus, an ancient Athenian and later Roman colony in Asia Minor (http://miletus.fhw.gr/). Similarly, examples of early immersive installations can be found in Austria (Ars Electronica, http://www.aec.at) and Japan (Intercommunication Centre, http://www.ntticc.or.jp/). These instances demonstrated the clear potential of emerging technologies for cultural heritage institutions, although their widespread adoption remained gradual and restrained.

Subsequently, the emergence of dedicated immersive reality museums, such as The Museum of Pure Forms (begun in 2002) and The Virtual Museum of Sculpture and Architecture (2008), signifies a notable shift in the adoption of XR technologies. Carrozzino and Bergamasco (2010) attribute this shift to the increasing acceptance and availability of similar technologies among the general public. The introduction of consumer devices like Nintendo's Wii (2006), featuring motion sensors, and the iPhone (2007), equipped with touch screens and camera-based applications, familiarized individuals with interactive technology, thereby reducing hesitancy toward immersive installations. The early virtual museums and exhibition spaces played a crucial role in shaping the current landscape of immersive experiences. Notably, these early ventures targeted the general public rather than requiring specialized training, unlike VR applications in fields such as medicine and sciences. Moreover, to ensure visitor flow within gallery spaces, the duration of museum experiences had to be relatively brief compared to more extended VR experiences developed by researchers like Salzman et al. (1999). This shift toward shorter durations and experiences tied to physical spaces paved the way for the emergence of fully virtual museums, exemplified by initiatives like The Exploratorium (1969) in Fig. 2.12, a public science museum, and The RePAIR project (2021), an EU-funded endeavor

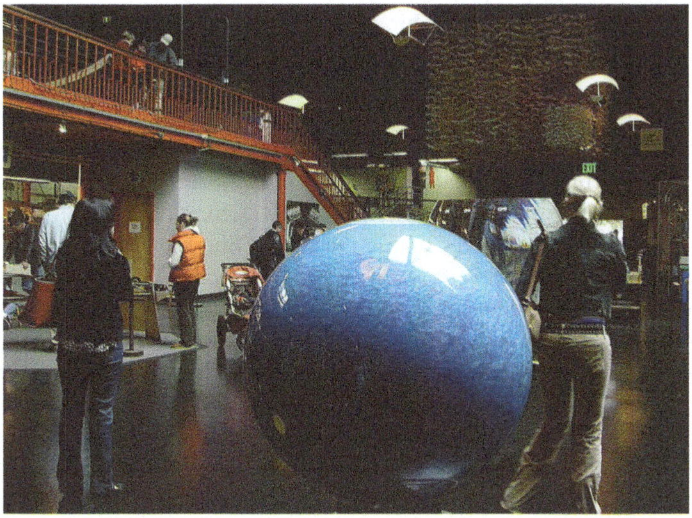

Fig. 2.12 Exploratorium, May 2008 (from Wikimedia Commons, licensed under CC-BY 2.0)

enabling users to reconstruct archaeological sites (https://cordis.europa.eu/project/id/964854). Consequently, the demand for user-friendly interfaces and immersive design, coupled with the need for concise experiences, has driven the design of VLEs that are widely embraced in contemporary contexts.

In the realm of virtual museums, the function of these digital spaces carries on the ritualization of space in the tradition of Camillo's Theatre of Memory where knowledge is organized in an exhibitionary and theatrical order. Carol Duncan provides insights into the performative and ceremonial aspects of museum architecture, spatial organization, and display in her article "The Art Museum as Ritual" (1995). Her analysis underscores the pivotal role of museums in fabricating and perpetuating dominant social ideologies and explores how viewers and users engage in civic rituals within the museum space. She argues that museum practices are deeply rooted in the construction and maintenance of these ideologies. However, amidst these societal underpinnings, the exhibitionary order of modern museums reveals an underlying spiritual dimension that arrived with the invention of aesthetics in the eighteenth century, which facilitated a transfer of spiritual values from the sacred to the secular realm of social life. The museum, as an institutional manifestation of this philosophical transference, endows art objects with transformative spiritual, ethical, and emotional power.

In the virtual museum landscape, the organization of knowledge takes on new dimensions in terms of place and presence. Digital spaces transcend physical limitations, enabling the creation of immersive environments that encompass diverse cultures and historical eras. Virtual museums serve as stages where rituals of exploration, discovery, and engagement unfold. Embracing the virtual realm allows museums to expand their roles as custodians of cultural heritage, reaching global audiences. Through interactive exhibits, virtual tours, and innovative storytelling techniques, these digital spaces provide transformative experiences that transcend geographical and temporal boundaries. While virtual museums inherit institutional practices and social ideologies from their physical counterparts, they also offer opportunities to reimagine the museum experience. The integration of aesthetics, technology, and interactivity in virtual spaces fosters new forms of engagement and interpretation. Visitors become active participants, navigating digital realms and shaping their own immersive experiences within the virtual museum's narrative.

In a similar fashion, the development of VLEs has gained widespread adoption as an effective approach to the teaching practices for arts and culture, encompassing both the digitization of existing museums and the creation of computer-generated replicas. The introduction of Google's Arts and Culture platform (https://artsandculture.google.com/) in 2011 marked a significant milestone, as it brought virtual museum visits into the mainstream and made them accessible to smartphone users. Google Cardboard further democratized the technology by introducing a head-mounted display that allows for immersive experiences at a lower cost. Various VLEs have been designed to facilitate tours of both real and virtual museums. Examples include the National Archaeological Museum of Marche in Ancona and the 3D reconstruction of the Roman Forum, where digitized photographs are transformed into virtual environments using the HTC Vive system and movement metrics of technique structure. With the HTC Vive system, users wear a VR headset that provides an immersive, three-dimensional visual experience. In addition to the headset, the system includes hand controllers that allow users to interact with the virtual environment. This interaction extends beyond passive observation; users can physically move within the virtual space and manipulate objects using the controllers. In the case of the 3D reconstruction of the Roman Forum, digitized photographs are transformed into a virtual environment that users can explore while wearing the HTC Vive headset. This means that users can move around within the reconstructed Forum, examining details, and interacting with elements of the environment using the hand-held controllers. It is a highly interactive and immersive experience that simulates the feeling of actually being present within the reconstructed historical site.

Additionally, the Rijksmuseum in Amsterdam and a virtual recreation of Gyeongju, South Korea (2015), have developed virtual environments specifically tailored for the Samsung Gear VR system. Free game engines like Unity3D (https://unity.com/) and Unreal Engine (https://www.unrealengine.com/en-US) have also been utilized to create virtual museum experiences for educational purposes. These engines serve two primary scenarios: firstly, enabling users with VR systems at home to explore virtual museum environments, effectively simulating physical visits. Here, individuals can navigate exhibits and access educational content, offering an immersive remote experience. Secondly, they enhance on-site museum visits by allowing visitors to use VR systems for various virtual experiences related to the art exhibits. This may include guided virtual tours, historical reenactments, or interactive simulations that provide deeper insights into the artworks and historical contexts. These technologies aim to augment the educational aspects of museum visits, offering engaging and immersive learning experiences accessible both remotely and on-site, catering to diverse audiences, from scholars to casual museum enthusiasts. Brennan and Christiansen (2018), for instance, have employed 3D scanners and high-quality photography to digitize the ancient sculpture collection in the Uffizi Galleries in Florence, importing them into Unity to construct an immersive learning environment. Similarly, the Universidad Nacional de San Agustin de Arequipa in Peru implemented a Unity-based experience for their art history students, demonstrating positive outcomes and student perceptions through an active learning strategy (Huaman et al. 2019).

The virtual museum serves as a prime example of how VR technology has been gamified to enhance the learning experience. In these virtual environments, students navigate through various rooms, earning points for each room visited and for actively engaging with the artworks on display. Survey data collected from students who participated in these virtual museum experiences not only demonstrated improved student outcomes but also highlighted the potential for even greater interactivity and engagement by allowing students to create their own exhibitions. Building on these findings, developers such as VIVE have capitalized on the results and introduced VIVE Arts (https://www.vivearts.com/), a platform that empowers users to both curate and showcase their own artworks within virtual spaces.

In recent years, there has been a growing emphasis on the digitization of collections and entire museums for viewing in simulated reality, facilitated by advancements in technology. Brown's (2020) edited volume on digital humanities and art history explores this trend, specifically addressing the emergence of virtual museums. Museums worldwide are actively engaging in the digitization process to make their collections more accessible to a wider audience. An exemplary initiative in this realm is the Center for the Art of East Asia at the University of Chicago, which has been at the forefront of developing digital technologies for archiving and viewing East Asian paintings and sculptures since 2006. Projects such as The Scroll Paintings Project (https://caea.uchicago.edu/projects/long-term-projects/digital-scrolling-paintings-project) and The Chinese Buddhist Caves Temple Project have been instrumental in enhancing access to art-historical resources and fostering collaboration and scholarship with artworks that are often physically inaccessible (Whitfield et al. 2000).

However, the pace of digitization is a significant challenge for museums to tackle individually. To expedite the process, many institutions are turning to service-oriented approaches, seeking assistance from third-party entities. Open-source technologies, including bespoke application programming interfaces (APIs), are being leveraged to enable "app communities" and other collaborators to recreate museum collections or curate new virtual museums. This collaborative model not only accelerates the digitization process but also encourages innovation and diversity in virtual museum creation, further expanding the scope of cultural heritage preservation and dissemination. These technologies empower developers to create modern museum apps with a wide range of capabilities that extend beyond the traditional "Guided Tour" paradigm. These capabilities often include interactive 3D visualizations of artworks AR experiences that overlay digital content onto physical exhibits, immersive VR tours, gamified learning modules, real-time location-based services within museums, access to curated multimedia content, and even social features that enable visitors to share their experiences and insights. These modern museum apps are evolving to offer enriched and personalized engagements, catering to the diverse preferences and expectations of museum-goers in the digital age. Next, we will delve into a noteworthy instance of a collaborative model that merits further discussion. While the aforementioned examples revolve around extant sites and cultural artifacts, an intriguing question arises: How can we approach the recreation of entities that have ceased to exist?

The Virtual Studiolo Project, developed by the Isabella d'Este Archive at IDEA, explores just that and the possibilities of leveraging immersive technologies to repatriate and reconstitute collections that are no longer extent. Inspired by the famed *studiolo* and *grotta* of Isabella d'Este (1474–1539) in sixteenth-century Mantua, the project aims to create an immersive, 3D environment that brings together the surviving objects and approximations of lost items in a virtual space (https://www.isabelladestearchive.org/virtual-studiolo). By utilizing VR, users can navigate and experience the meticulously recreated chambers, gaining insights into the cultural displays, collections, and personal retreats of the marchioness. Through this modern "remastering," the Virtual Studiolo seeks to inspire new research, inquiries, and deeper understanding of the structure and its collection, along with the cultural context that shaped it. The project is a collaborative effort between humanists, computer scientists, and digital artists.

At the same time, experiences have vastly expanded beyond even entire museums. The recreation of entire lost civilizations has also been carried out in projects such as Virtual Angkor at Monash University, Melbourne, Australia (https://www.virtualangkor.com/). As a collaborative effort between archaeologists, historians, and virtual history specialists, the project reconstructs the ancient city of Angkor in Cambodia. Through immersive VR and 360-degree videos, Virtual Angkor allows visitors to experience the bustling city during its peak in the thirteenth century. The project utilizes richly detailed 3D models and animated populations to create an immersive and historically accurate environment. By integrating archaeological and textual sources, the experience presents a holistic analysis of trade, diplomacy, climate, political power, and the sense of place in the medieval complex.

Other interactive examples include *Witness to Revolution* (2021) spearheaded by the University of Wisconsin-Stout (https://www.wttrgame.com/). This immersive VR game is set in Colonial Boston following the 1770 "Boston Massacre." Developed by the University of Wisconsin-Stout and Carleton College, the serious (or educational) game offers an historically accurate recreation of the period and aims to explore the impact of misinformation on the perception and interpretation of historical events. Players are tasked with investigating evidence, engaging with witnesses, and experiencing playable flashbacks of witness accounts as they navigate the complexities of contradictory information. By examining Paul Revere's print *The Bloody Massacre*, conversing with individuals, and sketching key objects, players develop their own interpretations of the events and strive to uncover the truth of what occurred on March 5, 1770.

Expanding virtual worlds have also experienced significant growth and transformation in recent years with the concept of an interoperable virtual world gaining widespread attention. The exploration of virtual worlds has opened up new possibilities for cultural heritage experiences, allowing users to delve into different time periods and locations. Initially, many virtual worlds emerged as massively multiplayer online role-playing games (MMORPGs), such as *EverQuest* and *World of Warcraft*, gradually evolving into expansive alternate realities that extended beyond gaming. One prominent example of a virtual world with cultural significance is Second Life (https://secondlife.com/), launched in 2003 (and still active). Second Life allowed users to interact (Fig. 2.13), create, and monetize their virtual creations, giving rise to a unique digital economy. Although widespread adoption of virtual worlds by mainstream businesses has yet to be realized, these developments have laid the groundwork for exploring the potential of virtual cultural heritage experiences (Sourmelis et al. 2017).

Another notable example is MindArk's Entropia Universe (https://www.entropiauniverse.com/), which initially began as a game in 2003 before expanding into a metaverse spanning multiple planets. Entropia Universe has broken several records for owning valuable digital properties, attracting significant attention and investment. The concept of purchasing virtual land has become a central focus in these virtual worlds. Decentraland (launched 2017), for instance, enables users to buy and sell virtual land parcels using cryptocurrency (https://decentraland.org/). The value of virtual land is influenced by factors such as location and the features present on the land, creating a complex economy within the virtual realm. However, the volatility of the crypto market introduces uncertainty, as the value of virtual land can fluctuate rapidly (Goanta 2020).

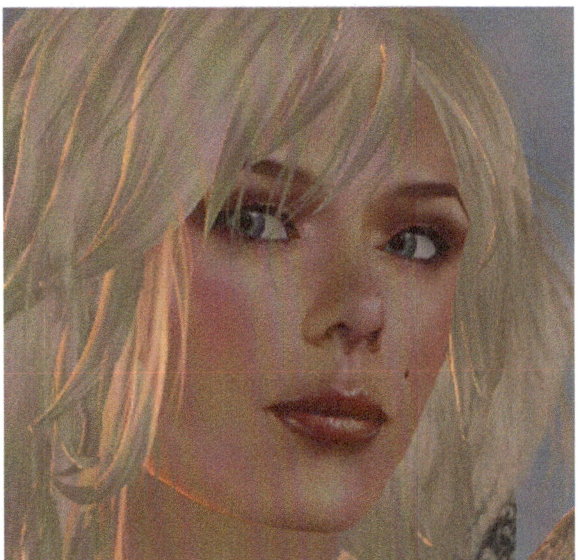

Fig. 2.13 Charlette Proto, Second Life (from Wikimedia Commons, licensed under CC-BY 3.0)

Virtual worlds and real estate prospecting within them present an intriguing confluence of opportunities and challenges in the context of cultural heritage preservation. The virtual domain offers the possibility of transcending geographical and temporal boundaries, thereby enabling users to engage with a wide array of cultural artifacts and historical periods. Yet, the question arises: do museums and cultural institutions view virtual real estate as an extension of their mission, or does this novel paradigm present insurmountable socio-economic barriers? Evidence suggests that museums and cultural institutions are indeed showing a growing interest in virtual platforms for disseminating knowledge and fostering engagement. Museums, for instance, have already begun to leverage VR for immersive exhibitions and interactive experiences that are inaccessible through traditional means. Virtual lands, therefore, can serve as the next evolutionary step in this trajectory. Moreover, digital platforms allow for scalability and reach that physical locations cannot match. Virtual lands can be accessed by a global audience, thereby offering an expanded geographical reach that aligns with the mission of many cultural institutions to disseminate knowledge broadly. Consequently, while upfront costs may exist, the long-term benefits in terms of reach and engagement may outweigh these initial limitations.

Thus, while the socio-economic barriers should not be underestimated, they do not necessarily preclude the effective use of virtual worlds for cultural heritage preservation. The dichotomy between virtual and physical spaces need not be an either/or proposition; rather, a nuanced, integrative approach can optimize the benefits while mitigating the limitations. Virtual worlds hold transformative potential for cultural heritage, serving as a complement to physical spaces and artifacts. Through careful planning and equitable design, museums and cultural institutions can harness the power of virtual lands to enrich cultural understanding and broaden access to diverse audiences. Therefore, the application of virtual environments in cultural heritage demands a balanced consideration of its economic, social, and educational ramifications. With digital twins and models of world cultural heritage monuments, museums, along with entire virtual worlds, the phase currently underway is the creation of a complete digital twin of the Earth itself. In recent years, the development of digital twins of the Earth has gained significant momentum, driven by the need to combat climate change and preserve nature. The European Commission, in collaboration with various organizations, has launched the Destination Earth initiative, aiming to create a highly accurate digital model of the Earth. This endeavor, supported by the Digital Europe Programme (2021–2027), seeks to monitor, model, and predict natural and human activity while exploring scenarios for sustainable development.

The advent of the DestinE system—a collaborative initiative encompassing contributions from the European Space Agency (ESA), the European Organisation for the Exploitation of Meteorological Satellites (EUMETSAT), and the European Centre for Medium-Range Weather Forecasts (ECMWF)—promises a paradigm shift in the utilization of space and environmental data. The system is structured around three core components, each of which serves a unique but interconnected role in the larger framework. Firstly, the Core Service Platform, operated by the ESA, serves as the nucleus for data provision and service orchestration. In computational terms, a Core Service Platform is a centralized hub that facilitates the delivery, management, and orchestration of services provided to end-users. It operates as the central point of access, ensuring that users can obtain necessary data and services in an integrated manner. The Core Service Platform streamlines the interaction between various data sources and services, thereby creating a cohesive environment for the utilization of complex space and environmental data.

Secondly, the concept of a Data Lake, managed by EUMETSAT, necessitates a more granular examination. In essence, a Data Lake is a storage repository that holds a vast amount of raw data in its native format until needed. Unlike traditional databases, where data is structured and indexed, Data Lakes allow for the storage of unstructured data, thereby providing immense flexibility for future uses. Within the DestinE system, the Data Lake serves as a flexible and scalable reservoir for storing meteorological data, thereby allowing for a wide range of analytical pursuits that extend beyond the original intent of the data collection (DelSignore 2022). These digital twins combine real-time observations and simulations to focus on weather-induced and geophysical hazards, as well as climate change adaptation.

The development of digital twins offers extensive capabilities for monitoring, analyzing, and forecasting environmental and cultural phenomena. For example, a digital twin focused on weather-induced and geophysical hazards could enable local and regional authorities to simulate and test strategies for mitigating floods, droughts, heatwaves, earthquakes, volcanic eruptions, and tsunamis. Another digital twin, dedicated to climate change adaptation, can supply crucial data and simulation tools that underpin efforts in sustainable agriculture, energy security, and biodiversity conservation. Beyond environmental applications, digital twins also promise significant benefits for cultural heritage preservation. As technology progresses, digital twins could be used to replicate and preserve cultural sites, artifacts, and historical landscapes, opening new possibilities for exploration, education, and research. This integration of digital representations of Earth's physical and cultural landscapes positions the virtual twin as a formidable instrument for understanding, valuing, and protecting our global heritage.

The preceding has provided an initial overview of virtual cultural heritage experiences, from their initial stages of limited adoption and technological constraints to the current era of accelerated progress. The past two decades have witnessed the gradual removal of barriers, including technological limitations, financial constraints, and temporal constraints, which have impeded the widespread development of immersive cultural heritage experiences. However, with the rapid advancement of generative AI text-to-image and video capabilities, a new frontier has emerged, rendering the laborious methods of photogrammetric techniques or 3D modeling software obsolete.

The rapid advancement of AI-driven technologies has unlocked boundless possibilities, revolutionizing the creation of highly accurate digital representations of various subjects. This breakthrough has ushered in a new era of immersive and captivating cultural heritage experiences that were once unimaginable. As we embark on the next chapter, it is essential to delve deeper into the profound impact of AI on the art world and explore the exciting prospects that lie ahead for immersive encounters with cultural heritage. The transformative power of AI is expanding the horizons of what can be achieved in virtual environments at an unprecedented pace. Novel techniques and tools are poised to revolutionize how we engage with and appreciate cultural heritage. From meticulously rendered digital replicas of historical artifacts and architectural marvels to interactive storytelling experiences that transport us to distant eras, AI-driven advancements offer unparalleled opportunities for exploration, education, and preservation. In the upcoming chapter, the potential implications of AI on the art world shall be discussed, shedding light on how these technologies are reshaping the creative process, curatorial practices, and audience engagement. Additionally, the implications of AI on the preservation and accessibility of cultural heritage shall be considered.

References

Bernheimer R (1956) Theatrum mundi. Art Bull 38(4):225–247
Bohern H (1969) A special copy of G.'s Theatrum mundi. In: The book collector, vol 18, no 1, p 92s
Brennan M, Christiansen L (2018) Virtual materiality: a virtual reality framework for the analysis and visualization of cultural heritage 3D models. In: Digital heritage 2018, 3rd edn
Brown K (ed) (2020) The Routledge companion to digital humanities and art history. Routledge
Bühnemann G (2003) Maònòdalas and yantras in the Hindu traditions, vol 18. Brill
Byrskog S (2022) Story as history-history as story: the gospel tradition in the context of ancient oral history. Brill
Carrozzino M, Bergamasco M (2010) Beyond virtual museums: experiencing immersive virtual reality in real museums. J Cult Herit 11(4):452–458
Cornford FM (1997) Plato's cosmology: the Timaeus of Plato. Lib. of Liberal Arts, NY (Cambridge, 1937)
Das A, Patnayaka R (2020) Role of axis, symmetry and fractal geometry as design parameters in extrusion of Hindu temple form and construction management. Constr Mater Manage 217
DelSignore P (2022) The new age of creative AI began in 2022. Medium, 23 Dec 2022. https://medium.com/predict/the-new-age-of-creative-ai-began-in-2022-ece07bb93350
Duncan C (1995) The art museum as ritual. Art Bull 77(1):10
Ficino M (1989) Three books on life. In: Kaske CV, Clark JR (ed and trans)
Ficino M, Jayne SR (1985) Commentary on Plato's symposium on love. Spring, Dallas
Geravand S, Mirzaei S (2020) Importance of mandala and its epistemological role in travelling in the Vajrayana. J Subcont Res 12(38):195–214
Goanta C (2020) Selling LAND in Decentraland: the regime of non-fungible tokens on the Ethereum blockchain under the digital content directive. In: Disruptive technology, legal innovation, and the future of real estate, pp 139–154
Grafton A (1999) Cardano's cosmos: the worlds and works of a Renaissance astrologer. Harvard University Press
Gupta R (2023) 90% of data in the world was generated over the last 2 years. Forbes. https://inshorts.com/en/news/90-of-data-in-the-world-was-generated-over-the-last-2-years-forbes-1572924629009
Hahm D (1977) Origins of Stoic cosmology. The Ohio State University Press
Hess J (1967) On some celestial maps and globes of the sixteenth century. J Warburg Courtauld Inst 30(1):406–409
Huaman EMR, Aceituno RGA, Sharhorodska O (2019) Application of virtual reality and gamification in the teaching of art history. In: Learning and collaboration technologies. Ubiquitous and virtual environments for learning and collaboration: 6th international conference, LCT

2019, held as part of the 21st HCI international conference, HCII 2019, proceedings, part II 21, Orlando, FL, 26–31 July 2019. Springer International Publishing, pp 220–229
Hutson J (2010) Renaissance proportion theory and cosmology: Giovanni Paolo Gallucci's Della simmetria and Dürerian neoplatonism. Stor dell'arte 125
Jones EG (2020) Right romance: heroic subjectivity and elect community in seventeenth-century England. Penn State Press
Jouanna J (2012) Galen's reading of the Hippocratic treatise the nature of man: the foundations of Hippocratism in Galen. In: Greek medicine from Hippocrates to Galen. Brill, pp 313–333
Kowal K (2019) The Borobudur temple: the Buddhist architecture in Indonesia. Bud Archit 18(2):5–19
Kubler G (2008) The shape of time: remarks on the history of things. Yale University Press
Kumar S (2021) Circumambulation in Indian pilgrimage: meaning and manifestation. Int J Sci Eng Res 12(1):232–243
Lomazzo GP (1973) Idea del tempio della pittura [Milan, 1590]. In: Scritti sull'arte (ed). Roberto Ciardi, Florence
Lowery Z (ed) (2015) Historiography. Encyclopaedia Britannica
Ma MY (2020) Mandalas in the making: the visual culture of esoteric Buddhism at Dunhuang, by Michelle C. Wang
Morgan MH, Warren HL (1914) Vitruvius: the ten books on architecture. Cambridge
O'Neal H (2020) Word embodied: the jeweled pagoda mandalas in Japanese Buddhist art, vol 412. Brill
Rodríguez-Ortega N (2019) Digital art history: the questions that need to be asked. Vis Resour 35(1–2):6–20
Roussou M (2001) Immersive interactive virtual reality in the museum. In: Proceedings of TiLE (trends in leisure entertainment)
Salzman MC, Dede C, Loftin RB, Chen J (1999) A model for understanding how virtual reality aids complex conceptual learning. Presence Teleop Virtual Environ 8(3):293–316
Schroeder R (1993) Virtual reality in the real world: history, applications and projections. Futures 25(9):963–973
Schuler RM (1980) Some spiritual alchemies of seventeenth-century England. J Hist Ideas 41(2):293–318
Sourmelis T, Ioannou A, Zaphiris P (2017) Massively multiplayer online role playing games (MMORPGs) and the 21st century skills: a comprehensive research review from 2010 to 2016. Comput Hum Behav 67:41–48
Tatarkiewicz W (1972) The great theory of beauty and its decline. J Aesthet Art Critic 31(2):165–180
Thorndike L (1923) A history of magic and experimental science: fourteenth and fifteenth centuries, vols 3 & 4. Columbia University Press
Tumadi NH, Ritonga H, Hasan H, Kirin A (2021) The concept of Syariat teaching in Samawi religion in the works of The Story of Civilization by Will Durant. J Techno-Soc 13(1):23–28
White H (2005) Introduction: historical fiction, fictional history, and historical reality. Rethink Hist 9(2–3):147–157
Whitfield R, Whitfield S, Agnew N (2000) Cave temples of Mogao: art and history on the Silk Road, vol 4. Getty Publications
Winfield PD (2019) The philosophy of the mandala. In: The Dao companion to Japanese Buddhist philosophy, pp 235–253
Yates FA (1966) The art of memory. The University of Chicago Press, Chicago, IL
Zaluchu S, Widjaja F (2019) A phenomenology approach: religious significance of mandala in ancient Nias Culture. In: Proceedings of the 1st seminar and workshop on research design, for education, social science, arts, and humanities, SEWORD FRESSH 2019, Surakarta, Central Java, Indonesia, 27 Apr 2019

Part II
Place and Immersion

Chapter 3
Art in the Age of Virtual Reproduction

Abstract The chapter addresses how modern technologies such as digital imaging, 3D printing, and virtual reality have revolutionized the ways in which art is experienced and engaged with audiences worldwide. It elaborates on the concept of "digital alchemy," where technology acts as a transformative agent, allowing not only for the reproduction of existing artworks but also for the creation of new forms of art that challenge traditional boundaries. This chapter examines the democratization of art through digital means, which has made it accessible to a broader public beyond the confines of galleries and museums. It also discusses the impact of these technologies on the preservation of cultural heritage, enabling the restoration and public display of artworks that were previously inaccessible or deteriorated. The narrative weaves through historical precedents set by mechanical reproduction and moves toward current trends where digital tools empower both creators and consumers of art. By offering a comprehensive analysis of the shift from physical to virtual in the art world, the chapter highlights the changing paradigms in how art is perceived and valued in contemporary society.

3.1 Digital Alchemy: Technology as Transformative Agent in Art and Culture

As the art and culture landscape enters a new era, the integration of artificial intelligence (AI) and machine learning (ML) technologies is revolutionizing the abilities of creatives and cultural heritage institutions alike. In recent years, the advent of generative AI art tools, such as Stable Diffusion, DALL-E 2, and Midjourney, has captivated the public's imagination while igniting intense discussions among artists, scholars, and art historians (Ansari 2022). These groundbreaking tools have raised profound questions about the future of creativity, the evolving relationship between artists and AI, and the ethical, legal, and social implications of AI-generated art, evoking a mix of excitement and apprehension among those involved in the creative process. However, it is crucial to recognize the transformative potential of these tools, offering cultural institutions of all sizes and resources the unprecedented ability to

J. Hutson, *Art and Culture in the Multiverse of Metaverses*, Springer Series on Cultural Computing, https://doi.org/10.1007/978-3-031-66320-8_3

create immersive experiences that were previously unimaginable. The following will explore the implications of generative AI for immersive cultural experiences, examining the evolving landscape of AI-generated art, its impact on artistic practice, and the profound questions it raises for the art world.

Generative AI art encompasses the utilization of AI and ML algorithms to create visual artwork, 3D models, videos, and more. These algorithms undergo training using extensive datasets, often obtained from Internet sources, to grasp patterns, styles, and visual components inherent in human-generated art. By comprehending these elements, generative AI models possess the ability to generate original imagery based on user inputs, such as text prompts or specific visual cues. In the case of text-to-image generation, the AI model takes a text prompt provided by the user and transforms it into a cohesive visual representation, employing the assimilated visual elements to produce a distinctive artwork. This process involves intricate interactions between AI algorithms and neural networks, facilitating the model's comprehension and interpretation of both the textual input and the visual output. It is crucial to note that the creative process diverges significantly from digital imaging using software tools like Adobe Photoshop and Illustrator. These applications, which are designed for graphic design and digital art, furnish users with an extensive range of tools, brushes, and filters to facilitate image creation, editing, and manipulation. Serving as sophisticated digital canvases and workspaces, these programs serve as platforms where artists and designers have honed their skills, techniques, and creativity to achieve their desired visual outcomes (Hutson et al. 2023).

In the realm of 3D computer graphics, specialized software is utilized to create mathematical representations of objects in three-dimensional space through a process known as 3D modeling. This process involves manipulating edges, vertices, and polygons to construct a digital representation of a physical body composed of data points connected by various geometric entities, such as triangles and curved surfaces, allowing for a realistic portrayal of the object (Camba et al. 2023). Texture mapping can also be applied to further define the surface of the model. The resulting 3D models can be displayed as two-dimensional images through a process called 3D rendering or used in computer simulations to simulate physical phenomena. The creation of these digital representations can be accomplished manually, algorithmically, or through scanning techniques. Manual modeling involves a process similar to sculpting, where geometric data is meticulously prepared to achieve the desired shape and form. To facilitate the creation of these volumetric models, specialized software known as 3D modeling software is employed. These software applications, such as Blender, Maya, 3DS Max, Adobe Dimension, Cinema 4D, and Adobe Substance 3D Sampler, enable artists and designers to construct and manipulate 3D models with precision and control (Metzer et al. 2023).

Similar to Adobe Photoshop and Illustrator, becoming proficient in 3D modeling software requires considerable time and dedication to refine the necessary skills to produce desirable results. The process of creating complex 3D models can be laborious and time-consuming, often taking hours to create a single object, even when starting with preexisting scans or data. To achieve proficiency in 3D modeling, individuals must familiarize themselves with the intricacies of the software, learning how to manipulate edges, vertices, and polygons to accurately represent the desired object in a three-dimensional space. The process requires an understanding of various modeling techniques, such as box modeling, sculpting, or procedural modeling, and becoming proficient in texture mapping and other surface definition techniques (Roughley and Liu 2022). Furthermore, 3D modeling artists must possess a keen eye for detail and a strong understanding of form, light, and composition to create visually compelling and realistic models. The iterative nature of the process often requires experimentation, refinement, and constant adjustment to achieve the desired outcome.

On the other hand, and in contrast to traditional digital imaging and modeling approaches, generative AI art operates autonomously or semi-autonomously generating visual content based on user inputs, such as text prompts or visual cues. The creative process in generative AI art is rooted in the tool's comprehension of artistic elements, style, and composition, acquired through training on relevant datasets. Unlike the hands-on, skill-based nature of digital imaging and modeling, where artists and designers directly manipulate visuals to achieve desired outcomes, generative AI art entails an emergent process. Here, the algorithm interprets the input or "prompt" of a user and generates visual content based on its acquired knowledge, establishing a more collaborative dynamic between the human user and the machine (Hutson and Harper-Nichols 2023).

The accessibility of these emerging AI tools has been greatly improved through their browser-based interfaces, eliminating the need for proprietary software and reducing barriers to entry. This approach enhances user-friendliness and promotes widespread adoption among artists and creators. The previously mentioned digital imaging and 3D modeling software often come with a substantial cost, making them inaccessible to some artists and designers. Additionally, these software applications typically require specialized equipment, such as powerful graphics processing units (GPUs), to run smoothly and handle the complex computations involved in rendering and manipulating high-resolution images and models. Furthermore, the installation of such software often necessitates a significant amount of storage space on a personal computer, along with additional enterprise and administrative rites for institutional units. These factors contribute to the financial and technical limitations associated with using traditional digital imaging and 3D modeling tools.

Instead of relying on real-time rendering, the advent of generative AI art has introduced cloud-based platforms that enable content generation. These platforms leverage the computational power and scalability of cloud computing to perform complex AI algorithms and generate visual content. This shift from local processing to cloud-based solutions eliminates the need for specialized hardware and allows artists and designers to access generative AI tools and create artwork using remote servers. By offloading the computational burden to the cloud, artists can work with larger datasets and benefit from the collaborative nature of cloud platforms, where multiple users can contribute to the generation process simultaneously. This cloud-based approach expands the accessibility of generative AI art by removing the hardware constraints and providing a more flexible and collaborative environment for artistic exploration.

The user-friendly interface and quick training setup of generative AI tools have resulted in their widespread and rapid adoption, making them one of the fastest-adopted technologies in history. However, this swift acceptance of AI-driven art has also ignited a substantial backlash from creatives who have followed traditional training methods. Within the artistic and design communities, discussions have emerged surrounding issues such as copyright infringement, as well as debates about the authenticity and classification of AI-generated art as a distinct genre. In a significant legal development, a US court in Washington, D.C., ruled that a work of art generated solely by artificial intelligence could not be copyrighted under US law. The court upheld the US Copyright Office's rejection of an application filed by computer scientist Stephen Thaler on behalf of his artificial intelligence system known as Device for the Autonomous Bootstrapping of Unified Sentience (DABUS). This landmark decision resonates with another instance wherein the Copyright Office rejected an artist's claim for copyrights on images created through an AI system called Midjourney. The artist had contended that the AI system was integral to their creative process (Brittain 2023). These concerns underscore the intricate and dynamic connection between AI-driven creativity and traditional artistic practices, emphasizing the necessity for ongoing exploration and dialogue within the art world. As technological innovation continues to push the boundaries of artistic expression, there is a perpetual need to revisit and reevaluate the relationship between art, society, and creativity.

Throughout history, the art world has witnessed profound disruptions driven by emerging technologies. These disruptions have challenged established norms, reshaped artistic practices, and influenced cultural significance. Each paradigm shift, marked by transformative moments in technology, has introduced new techniques, materials, and mediums, while also sparking critical discussions about the social and historical implications of art. By examining these disruptions, a deeper understanding of the dynamic relationship between art and technology can be gained, along with how it has shaped the current artistic landscape. The introduction of printmaking and the printing press around 1440 marked a significant milestone in the intersection of art and technology. This technological advancement enabled the mass production of images and texts, fundamentally transforming the dissemination of knowledge and challenging the authority of handwritten manuscripts (Burke 2021). The accessibility and affordability of printed materials democratized access to art, expanding its reach to broader audiences. Consequently, the advent of printmaking sparked debates and discussions surrounding the originality and authenticity of artworks. The transition from individually crafted, unique pieces to mechanically reproduced prints disrupted established notions of artistic creation and ownership, leading to a reevaluation of the role of the artist and the value of artistic production.

The emergence of photography and photomechanical processes in the nineteenth century introduced a transformative wave in the realm of art. This groundbreaking medium disrupted the long-standing dominance of traditional painting and portraiture, presenting a new method for capturing and portraying the world. The capacity of the new medium to faithfully reproduce reality with precision and accuracy sparked profound discussions concerning the essence of art, the role of the artist, and the dynamic interplay between art and technology. Moreover, photography revolutionized temporal perception, memory construction, and the documentation of historical events, reshaping how society remembers and interprets the past (Schwartz and Ryan 2021).

Along the same lines, the emergence of computer-generated imagery and digital art in the 1960s signified a notable upheaval within the art world. Through the utilization of computers and digital technologies, artists gained access to expanded avenues of creative expression, enabling them to push the boundaries of traditional artistic practices. Digital art disrupted conventional categorizations of art forms, engendering a blending of mediums and technologies. It prompted contemplation regarding the singularity of artworks, the role of artists as programmers or curators of algorithms, and the evolving dynamic between viewers and the artwork itself (Du Preez 2020). In each instance of these creative shifts, the introduction of emerging technologies has heralded a wave of new tools and techniques that have simultaneously enhanced and challenged the art world. These disruptions have exerted a profound influence not only on the act of artistic creation itself but also on the perception, valuation, and consumption of art. They have ignited discussions concerning the fundamental nature of art, the evolving role of artists, and the profound effects of technology on the creative process. By delving into these pivotal moments in history, we can glean invaluable insights into the intricate interplay between art and technology, shedding light on the ongoing AI-driven evolution of the art world.

3.2 From Manuscripts to the Masses: Democratizing Knowledge and Art

The emergence of the printing press, invented by Johannes Gutenberg in 1440, brought about a profound revolution in the realms of art and information dissemination. This revolutionary technological development facilitated the mass production of books and artworks, effectively democratizing access to knowledge and creativity (Füssel 2020). The advent of relief printing, notably the technique of woodblock printing, can indeed be traced back to East Asia and its transformative impact on the dissemination and mass production of visual artifacts is worthy of scholarly attention. Originating in China and later adopted expansively in Japan and Korea, woodblock printing served as a revolutionary medium that drastically influenced not merely the reproduction of textual content but also the realms of visual art and cultural artifacts. One of the most illustrious examples that elucidates the profound influence of woodblock printing in East Asia is the compilation of the Chinese Buddhist canon, often referred to as the Tripitaka. Created during the Tang dynasty, the Tripitaka was an enormous endeavor that entailed the carving of thousands of woodblocks for printing sutras and commentaries. The accessibility and affordability of these printed texts revolutionized religious practice and scholastic endeavors, making Buddhist teachings widely available across different strata of society (Chia 2020).

Printmaking, as an artistic practice, flourished in this era in the West with notable artists such as Albrecht Dürer (1471–1528) employing intricate woodcut and engraving techniques to produce visually captivating prints. These techniques not only enabled artists to reproduce their works on a larger scale but also facilitated the rapid dissemination of artistic ideas, transcending geographical boundaries and captivating audiences across continents. The availability of mass-produced prints facilitated Dürer's reach to a broader market and resulted in significant sales. A notable example of his success is his highly popular *Apocalypse* series, which he began after settling in Nuremberg in 1497. During this period, Dürer shifted his focus primarily to printmaking as a more reliable source of income compared to pursuing painting commissions (Low 2019). Alongside his detailed depiction of the apocalyptic events, he undertook an ambitious and innovative project to create an illustrated edition of the Passion of Christ, working on it concurrently. Dürer personally executed the preparatory drawings and produced the woodblocks used for printing both the images and accompanying text. Similar to his approach in the *Apocalypse* series, the illustrations in the *Passion* series were presented as full-page works accompanied by relevant Biblical verses, enabling viewers to seamlessly experience the scenes through visual and textual narratives. The series holds great significance in the art market's history as it capitalized on limited-edition series and catered to different clienteles, with larger and more expensive engravings appealing to wealthier patrons and smaller woodcuts catering to lower-income consumers. The combination of the intricate and detailed nature of engraving, coupled with the mass production and commercial success of prints, brought about a revolution in the art market during the Renaissance.

Simultaneously, the proliferation of mechanically reproduced texts and images played a pivotal role in disseminating ideas during a period of profound social and religious transformation. The Protestant Reformation, ignited by Martin Luther's Ninety-Five Theses in 1517, challenged the authority of the Roman Catholic Church and disrupted established religious doctrines (Hendrix 2019). The availability of affordable printed materials, including engraved prints, facilitated the swift dissemination of reformist ideas and allowed them to reach a wide audience, contributing to the momentum of the Reformation. The speed and widespread distribution made possible by printmaking positioned it as a potent instrument for disseminating propaganda, shaping public opinion, and fostering intellectual discourse.

Printmaking not only served as a means to disseminate religious and secular imagery but also played a crucial role in sharing scientific advancements and ideas across regions and continents. The medium facilitated the rapid spread of anatomical, mathematical, geometrical, and astronomical illustrations, contributing to the progress of various fields of study. One notable example is the work of Andreas Vesalius (1514–1564), the preeminent anatomist of the sixteenth century, whose highly influential anatomical treatise, *De Humani Corporis Fabrica* (On the Structure of the Human Body), first published in 1543 and seen in Fig. 3.1, revolutionized the understanding of human anatomy. The tome featured the most detailed and accurate illustrations of the human body to date, showcasing the anatomist's meticulous dissections and observations. These anatomical prints not only advanced medical knowledge but also influenced the practices of physicians across Europe (van der Wal 2020).

During the Enlightenment era, a significant endeavor to compile and systematize human knowledge took shape through the publication of comprehensive works, such as general encyclopedias. One prominent example is *Encyclopédie, ou dictionnaire raisonné des sciences, des arts et des métiers* (Encyclopedia, or a Systematic Dictionary of the Sciences, Arts, and Crafts), edited by Jean le Rond d'Alembert

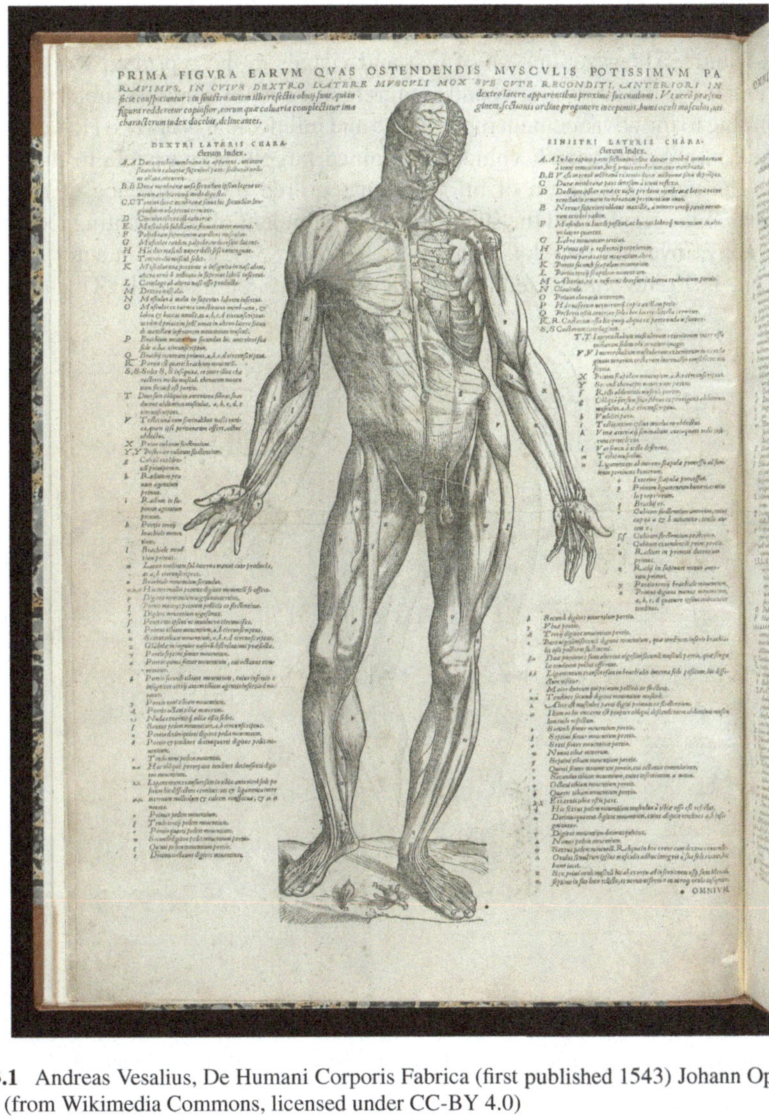

Fig. 3.1 Andreas Vesalius, De Humani Corporis Fabrica (first published 1543) Johann Oporinus, Basel (from Wikimedia Commons, licensed under CC-BY 4.0)

(1717–1783). Spanning from 1751 to 1772, this monumental publication encompassed a vast range of subjects and featured a multitude of engraved illustrations that accompanied the text. The illustrations included in this work depicted various scientific principles, mathematical concepts, as well as machinery, screws, and tools, among other subjects like glassware and bottles seen in Fig. 3.2. Its comprehensive text, accompanied by these detailed illustrations, provided readers with a wealth of information across diverse fields of study. The engravings, in particular, served as visual aids that facilitated the understanding and communication of complex ideas and concepts (Vigier et al. 2020).

The advent of printed illustrations revolutionized the sharing of knowledge and facilitated intellectual exchange among scholars, scientists, and artists during various fields of study. The combination of text and visuals in these printed works significantly enhanced the accessibility and comprehensibility of complex subjects, thereby contributing to the advancement of scientific and intellectual progress. However, until the nineteenth century, such volumes remained largely inaccessible to the general public. The democratization of information required the emergence of two more inventions: lithography and the subsequent development of new printing techniques. Lithography, invented by Alois Senefelder (1771–1834), a German actor and playwright, in 1796, revolutionized the production of printed materials. This technique

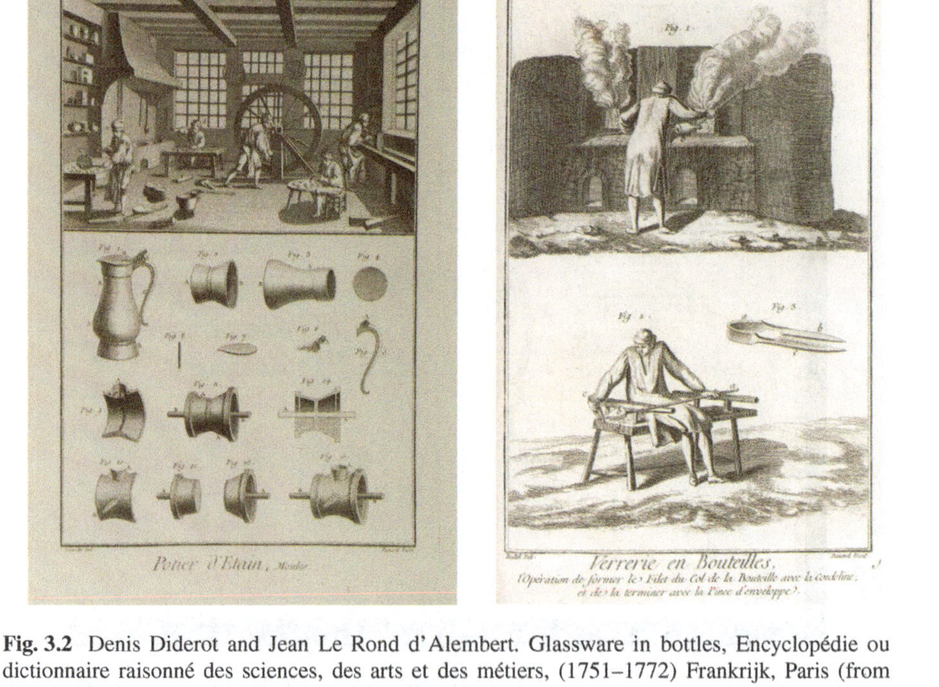

Fig. 3.2 Denis Diderot and Jean Le Rond d'Alembert. Glassware in bottles, Encyclopédie ou dictionnaire raisonné des sciences, des arts et des métiers, (1751–1772) Frankrijk, Paris (from Wikimedia Commons, licensed under CC0)

allowed artists to draw directly on a stone or metal plate using a greasy crayon or ink [some missing steps] and then transfer the image onto paper. Lithography provided artists with greater artistic freedom, enabling them to create spontaneous and expressive marks on the printing surface. This not only contributed to the rise of printmaking as a distinct art form but also facilitated the illustration and dissemination of new printed materials to a broader audience (Sabour 2021). The invention of lithography opened up new possibilities for artistic expression and expanded the range of printed materials available to the general public.

The development of lithography was also followed by advancements in printing techniques that further broadened the accessibility of printed materials. Innovations such as steam-powered printing presses and the use of mechanized typesetting, such as the Linotype machine, increased the speed and efficiency of the printing process. These technological advancements significantly reduced production costs and enabled the mass production of printed materials, making them more affordable and widely available to a larger segment of society (Singh 2018). Initially consisting primarily of text, these newspapers—including prominent publications like *The Times* (originally known as *The Daily Universal Register*, founded in 1785) in London and *Le Figaro* (founded in 1826) in Paris—later embraced the integration of printed illustrations, marking a significant development in the dissemination of ideas and the emergence of popular culture.The inclusion of visuals in these publications allowed for the mass production and widespread distribution of news, opinions, and entertainment. The synergy between text and visuals not only enhanced the overall reading experience but also enabled newspapers to engage readers through visual storytelling and informative imagery. Illustrations became a powerful tool in capturing the attention of readers, conveying information, and evoking emotional responses to shape public opinion (Aali 2021).

During the July Monarchy in France (1830–1848), a significant transformation occurred in the realm of printmaking and information dissemination, driven by the rise of illustrated newspapers and magazines. This period saw the emergence of influential publications like *Le Charivari* (1832–1937) and *La Caricature* (1830–1843) that utilized lithography as a medium to produce satirical and political cartoons. These illustrated periodicals played a crucial role in influencing public opinion and critiquing the social and political climate of the time, employing the skillful lithographs of artists such as Honoré Daumier (1808–1879), as seen in his caricature in Fig. 3.3 (Duncan-O'Neill 2021). In all, the impact of the lithographic revolution extended beyond the realm of art and popular culture. It laid the groundwork for the modern media landscape by setting the stage for further advancements in mass communication and the integration of images and text. The convergence of printmaking, popular culture, and technological innovations during this period shaped the trajectory of media and communication, leaving a lasting influence on the development of visual media.

The advancements in printmaking throughout history have left an indelible mark on artistic expression, the exchange of ideas, and cultural dissemination. These transformative developments have fundamentally altered the accessibility of knowledge and expanded the reach of artists, fostering intellectual and artistic movements of

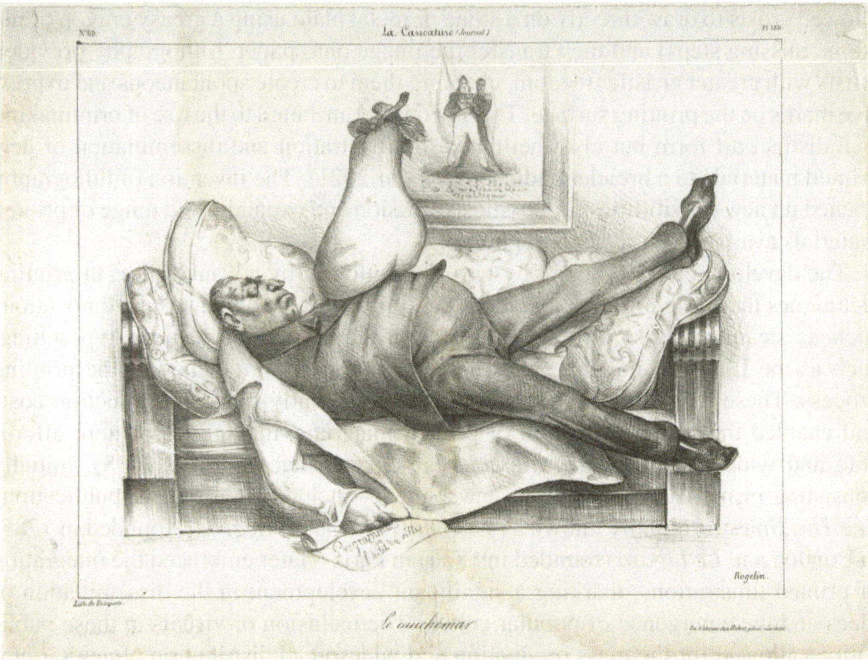

Fig. 3.3 Honoré Daumier, Le Cauchemar (The Nightmare), lithograph, La Caricature no. 69, February 23, 1832 (from Wikimedia Commons, licensed under CC0)

great significance. The invention of the printing press and subsequent evolution of printmaking techniques played pivotal roles in shaping key periods such as the Renaissance and the Enlightenment, propelling artistic and intellectual exploration to new heights. However, the disruptive force of technology did not cease with printmaking alone, as the advent of photography brought about another seismic shift in the art world. With its ability to capture and reproduce visual phenomena with unprecedented precision, photography further revolutionized artistic expression, challenging traditional notions of representation and opening up new possibilities for artistic creation.

3.3 Beyond the Frame: Exploring the Artistic Boundaries of Photography

During the Industrial Age, the emergence of photography as a new pictorial medium brought about a profound transformation in the concept of art and the role of the artist. Photography, which debuted in the early nineteenth century, swiftly revolutionized the world of visual arts, pushing the boundaries of traditional representation. The birth of photography as a medium can be traced to the patenting of the daguerreotype in 1839 by Louis-Jacques-Mandé Daguerre (1787–1851). Thus, the eponymously named daguerreotype process involved the exposure of a silver-plated copper plate to light, resulting in a direct positive image that captured the appearance of the natural world with unprecedented precision (Daguerre 1839).

Despite its intricate and time-consuming nature, the daguerreotype process gained immense popularity due to its remarkable ability to produce highly detailed and permanent images. The chemical treatment and exposure times required careful precision, ranging from several seconds to several minutes. Consequently, daguerreotypes became treasured possessions for individuals who sought to immortalize their loved ones, commemorate significant events, or capture detailed landscapes. A prominent figure in the early days of photography, French photographer Félix Nadar (1820–1910), made a significant impact with his portraits. Nadar captured images of notable personalities of the time, including writers, artists, and politicians. His renowned portrait of the French author and poet Charles Baudelaire, taken in 1855 seen in Fig. 3.4, exemplifies the precision and level of detail achievable through the daguerreotype process (Paigneau 2019).

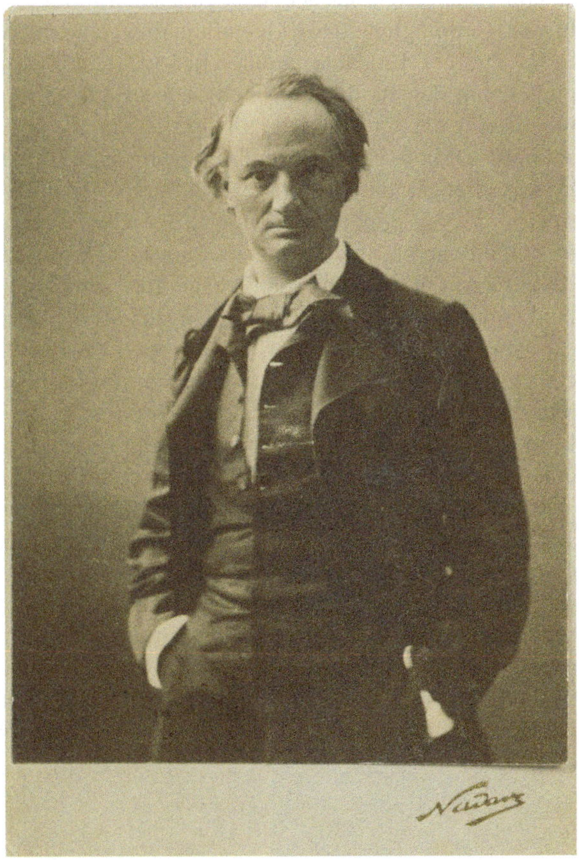

Fig. 3.4 Félix Nadar, Charles Baudelaire, 1855 (from Wikimedia Commons, licensed under CC0)

However, the limitations of daguerreotypes in terms of mass production hindered their adoption for illustrations in newspapers and other popular media. To overcome this constraint, a significant breakthrough came with the introduction of the wet plate collodion process, which revolutionized the field of photography by enabling greater ease and efficiency in image production. The technique, pioneered by Frederick Scott Archer (1813–1857) in 1851, involved coating a glass plate with a collodion solution and sensitizing it with silver nitrate, allowing for shorter exposure times. The wet plate collodion process quickly gained prominence and became the dominant photographic method throughout the mid-nineteenth century, providing photographers and artists with enhanced flexibility and expanded creative possibilities (Kontou et al. 2022).

The advent and widespread adoption of photography brought about a significant shift in the representation of various subjects, ranging from portraiture to landscapes and historical events. One noteworthy photographer who made a lasting impact during this period was Mathew Brady (1822–1896), often hailed as the "father of photojournalism." Brady's photographs, particularly those taken during the American Civil War, served as a seminal documentation of the realities of war and offered the public a fresh and unvarnished perspective on the conflict. His images provided an unprecedented visual account of the harsh conditions, devastation, and human cost of war (Coddington 2022). One of the most renowned photographic series by Brady is *The Dead of Antietam*, captured in 1862, and illustrated in Fig. 3.5. This haunting photograph depicts the aftermath of the Battle of Antietam, known as one of the bloodiest battles in American history. The image portrays a field strewn with fallen soldiers, evoking a somber and poignant representation of the human toll exacted by war. Alongside Brady's other photographs of the war, this image played a pivotal role in shaping public perception of the Civil War, offering a visual testimony to its brutality and far-reaching consequences (Hazard 2023). Simultaneously with the development of the daguerreotype, another significant breakthrough occurred in the field of photography in 1839. William Henry Fox Talbot (1800–1877) introduced the calotype process, also known as the talbotype, which revolutionized the medium and contributed to its democratization. In contrast to the daguerreotype's direct positive image on a copper plate, the calotype process involved the use of a paper negative from which multiple positive prints could be produced. This innovative technique made photography more accessible and reproducible (Lipton and Lipton 2021).

The calotype process employed a series of steps: sensitizing a sheet of paper with silver iodide, exposing it in a camera to capture the image, and subsequently developing it through chemical processes. The calotype offered significant advantages over the daguerreotype as it allowed for the creation of multiple prints from a single negative, providing a new degree of reproducibility. Moreover, the calotype produced images with a softer and more tonally varied aesthetic compared to the sharpness of the daguerreotype. These characteristics contributed to the democratization of photography, as the calotype process enabled the mass production and distribution of images.

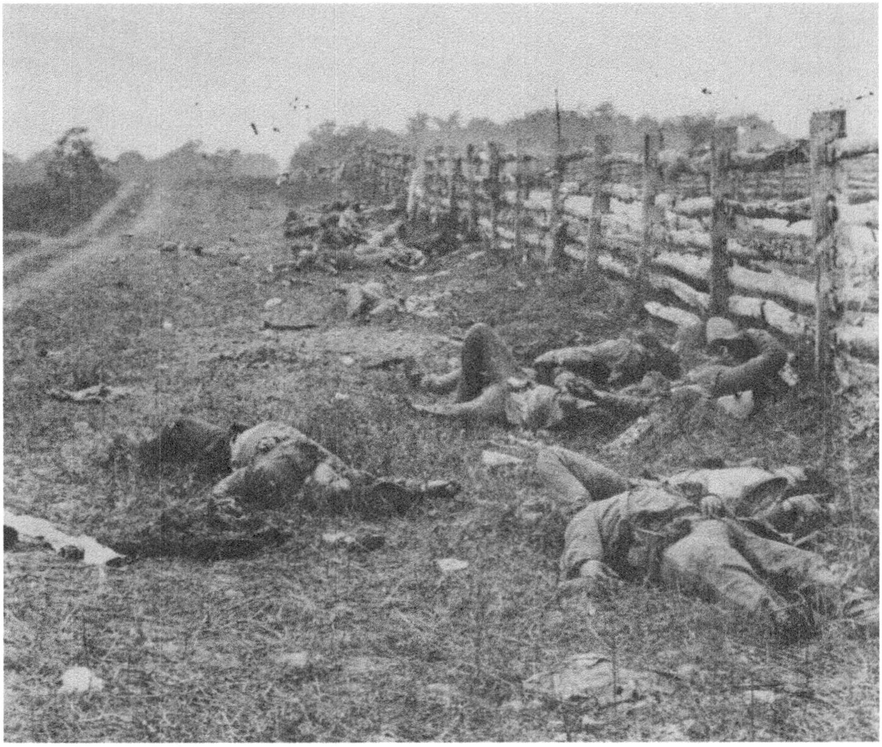

Fig. 3.5 Mathew Brady, The Dead of Antietam, 1862 (from Wikimedia Commons, licensed under CC0)

An exemplary illustration of the transformative impact of the calotype process is evident in the influential series by Talbot, *The Pencil of Nature* (1844–1846). One such example can be seen in Fig. 3.6, while the series also holds the distinction of being the first commercially published book illustrated with photographs. Through the publication, Talbot not only showcased the technical capabilities of the calotype process but also demonstrated its potential as a tool for visual documentation and artistic expression. The illustrated text encompassed a diverse range of subjects, including landscapes, architectural studies, still-lives, and portraits. The selection of subjects highlighted the versatility and artistic potential of the calotype process. By capturing the intricate details of nature, architectural structures, and everyday objects, Talbot revealed the aesthetic richness that could be achieved through photography, challenging the prevailing perception that it was merely a scientific tool, though many advances were also made in this area (Halkyard 2021).

At the same time, the advent of photography had a profound impact on the established art academies of Europe and the Americas. These academies, such as the Royal Academy of Arts in London (founded in 1768), the École des Beaux-Arts in Paris (established in 1648), and the National Academy of Design in New York (founded in

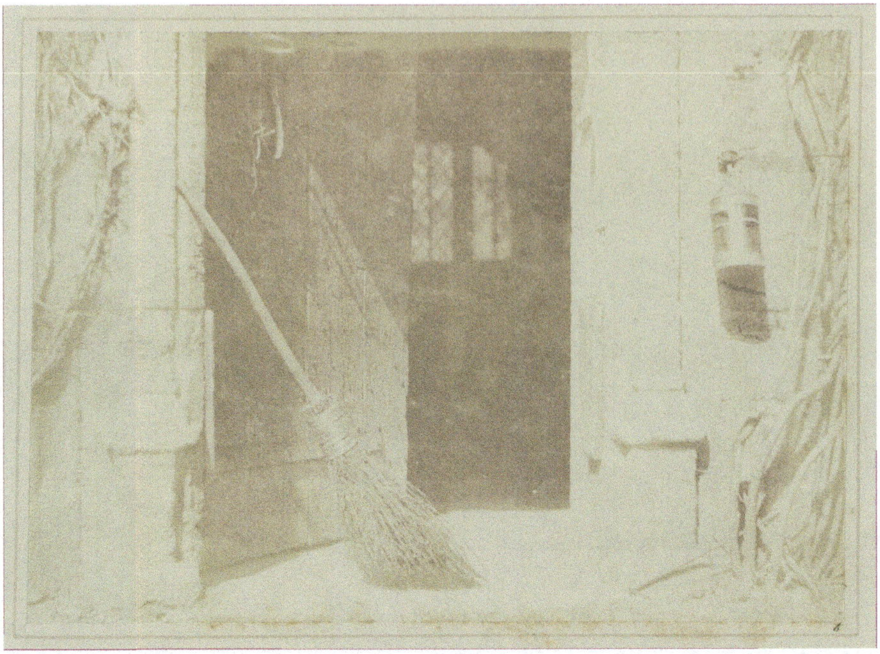

Fig. 3.6 William Henry Fox Talbot, The Pencil of Nature (1844–1846), calotype (from Wikimedia Commons, licensed under CC0)

1825), were influential institutions responsible for training and educating artists. The introduction of photography as a new medium forced these academies to confront the question of whether it should be recognized as a legitimate form of fine art and how it should be integrated into their curricula (Lee 2020b). Initially, the academies, including the Royal Academy of Arts and the École des Beaux-Arts, were skeptical of the fledgling medium's artistic merits and perceived it as a threat to the traditional disciplines being championed. They resisted accepting photography as a valid art form, dismissing it as inferior.

However, as photography gained recognition and popularity, the academies began to acknowledge its artistic potential and started incorporating it into their educational programs (Kremnitzer 2022). Across the Atlantic in the United States, the National Academy of Design played a significant role in shaping the reception and acceptance of photography as an art form. Like its European counterparts, the National Academy initially took a conservative stance toward photography. However, as the artistic merits of the medium became increasingly evident, the academy embraced photography and provided opportunities for photographers to exhibit their work alongside more traditional art forms. This recognition by prestigious institutions like the National Academy of Design contributed to the establishment of photography as a respected and integral part of the art world (Mccoy 1972). The inclusion of photography within the curriculum and exhibitions of these art academies marked a pivotal moment in the history of the medium. It reflected the changing landscape of the art world and the growing recognition of photography as a significant artistic form. These institutions played a crucial role in shaping the reception and acceptance of photography as an art form, facilitating its integration into the canon of fine art.

The art historical impact of the new medium is also evident by the 1930s. Walter Benjamin, for instance, provides insights and a cultural criticism perspective in his "The Work of Art in the Age of Mechanical Reproduction" (1939) (Benjamin 2008). The democratization of photography, driven by technological advancements, raises questions about the authenticity, "aura," and value of artworks. Benjamin discusses in terms of the concepts of the "cult value" and "exhibition value," which serve as valuable heuristics for understanding the metamorphoses in the valuation and significance of artworks in the context of technological innovation. "Cult value" refers to the intrinsic, often religious or ritualistic, importance attached to an artwork, which is not primarily intended for public exhibition. This form of value is rooted in the artwork's unique existence at a specific time and place, often cloaked in mystery and inaccessible to the general public. In contrast, "exhibition value" pertains to the utility of an artwork as an object for display, subject to public gaze and interpretation. As artworks increasingly become reproducible, especially with the advent of photography and other mechanical means, their exhibition value often supersedes their cult value. Therefore, Benjamin's examination of the tension between cult value and exhibition value in art sheds light on the changing landscape of artistic production, dissemination, and reception in the age of technological reproduction. Photography, as a reproducible and widely circulated medium, has the potential to transcend cultural and geographic boundaries, allowing for encounters and interventions in diverse social and political contexts, influencing public perception, and shaping collective memory.

Photography not only had a significant impact on academic curricula but also brought about a profound transformation in traditional art forms, particularly painting. The advent of photography challenged the purpose and role of traditional representational art by offering a precise and accurate means of capturing realistic images (Snyder and Allen 1975). The ability of the process to objectively depict the representational world posed a new challenge to established art institutions that had long upheld the principles of academic art. The emphasis on meticulous technique and faithful representation was called into question as photography demonstrated its capacity to faithfully reproduce reality. For instance, the influence of photography on art can be seen in the works of the Impressionist painters, including Claude Monet (1840–1926) and Pierre-Auguste Renoir (1841–1919), who drew inspiration from the fleeting effects of light and atmosphere that photography struggled to capture at the time. Other artists, such as Edgar Degas (1834–1917) and Édouard Manet (1832–1883), incorporated compositional strategies and cropping techniques utilized in photography, resulting in dynamic compositions and intimate portrayals of everyday moments (Staub 2019). The introduction of photography freed painting, and later sculpture, from the constraints of true-to-life representation, allowing artists to explore alternative forms of human experience and expression.

As photography continued to exert its influence on the realm of fine art, the nineteenth and twentieth centuries witnessed notable advancements that had a lasting impact on the visual arts. One significant development during this period was the introduction of tintypes, also known as ferrotypes, in the 1850s. Tintypes provided a cost-effective and durable option for portrait photography, making it more accessible to a wider range of individuals. The affordability and resilience of tintypes played a crucial role in democratizing the medium, allowing a greater number of people to participate in and enjoy the art of photography. Likewise, another significant breakthrough in the field of photography occurred with the invention of roll film by George Eastman (1854–1932) in 1888. This innovative technology marked a significant departure from the earlier use of glass plate negatives, transforming the photographic process. Roll film introduced efficiency and convenience for photographers, eliminating the need for cumbersome equipment and allowing for multiple exposures to be captured seamlessly. This advancement expanded the range of subjects and moments that photographers could capture, enhancing the artistic possibilities of the medium (Sturchio 2020). The introduction of tintypes and the invention of roll film exemplified the ongoing progress in photography, contributing to greater accessibility, efficiency, and creative freedom for artists. These advancements played a pivotal role in shaping the visual arts, influencing photographers' approach to their craft, and expanding the medium's potential for artistic expression.

A few decades later, the invention of color photography marked a significant milestone in the evolution of the medium. Initially, color photography was an intricate, laborious process, largely confined to specialized practitioners and institutions, predominantly in Western Europe and the United States. However, technological breakthroughs facilitated the democratization of this medium, extending its reach to diverse geographical locales and social milieus (Hirsch and Erf 2014). While early experiments with color had been conducted, it was the release of Kodachrome film by Kodak in 1935 that brought practical and widely accessible color photography to the forefront. Kodachrome offered a groundbreaking color reproduction process that utilized dye layers to capture and preserve colors with exceptional accuracy and vibrancy. This breakthrough provided photographers with a new dimension to explore—the realm of color—allowing for the investigation of rich interplays of hues, tones, and shades within their compositions (Pope 2020). With the advent of color photography, artists gained the ability to experiment with color relationships, infuse their works with distinct moods and atmospheres, and evoke emotions in ways that surpassed the limitations of black-and-white photography.

Advancements in color photography continued, leading to the development of various color processes and techniques. The emergence of color film and the accessibility of color printing enabled photographers to create vibrant and visually captivating images. Artists such as William Eggleston (1939–), Saul Leiter (1923–2013), and Ernst Haas (1921–1986) embraced color photography as a means of artistic expression, capturing the beauty and nuances of the world in vivid hues. The availability of color photography also influenced other visual art forms. Painters and other artists drew inspiration from color photographs, incorporating the vibrant palettes and dynamic compositions seen in photographs into their own works. The exploration of color and its impact on artistic representation became a central focus in many artistic movements and styles, further blurring the boundaries between photography and other visual arts. This cross-pollination of ideas and techniques enriched the artistic landscape and contributed to the ever-evolving nature of photography as an art form.

In all, the history of photography illuminates the profound impact of technological reproduction on the practice and perception of art. From the early daguerreotypes and calotypes to the advancements in roll film and color photography, each technological innovation expanded the reach and potential of photography. The introduction of tintypes democratized portrait photography, making it more affordable and accessible to a broader range of individuals. The invention of roll film brought efficiency and convenience, allowing photographers to capture multiple exposures and fleeting moments with ease. The emergence of color photography enriched the medium, enabling artists to explore the interplay of hues and tones, and infuse their works with vibrant expressions of the world. The advancements in photography not only transformed the medium itself but also had a profound influence on other art forms. Painters drew inspiration from the vibrant palettes and dynamic compositions seen in photographs, blurring the boundaries between photography and traditional visual arts. Photography's ability to faithfully reproduce reality challenged the purpose and techniques of representational art, leading to the exploration of new artistic styles and modes of expression.

3.4 Reel Transformations: The Advent of Film and Artistic Disruption

Having considered the transformative impacts of photography—ranging from its democratization to its influence on visual culture, representation, and the commodification of images—the discourse now turns toward another watershed moment in the chronology of art and technology: the invention of film. This new medium emerged not merely as an extension of photographic techniques, but as a revolutionary form of artistic expression and public engagement unto itself. Just as photography had redefined the art of capturing a moment, film expanded the artistic lexicon to include motion, narrative, and a multiplicity of viewpoints.

The Lumière brothers, Auguste (1862–1954) and Louis Lumière (1864–1948), presented the Cinématographe to the public in December of 1895, in the basement of the Grand Café on the Boulevard des Capucines in Paris. This moment is often cited as the birth of cinema as a mass medium (Allan 2008). Unlike Thomas Edison's (1847–1931) Kinetoscope—introduced in 1891, which was more of a peephole viewer for a solitary spectator—the Cinématographe was capable of projecting moving pictures onto a screen, making it accessible to a larger audience. In this sense, the Lumière brothers' invention transformed film from a solitary experience to a collective one, institutionalizing the social dimension of film from its nascency (Carlson and Gorman 1990).

Advancements did not stop at the technological level; they extended into the realm of narrative and thematic sophistication. One of the pioneers in this regard was D. W. Griffith (1875–1948), an American filmmaker, whose groundbreaking work, *The Birth of a Nation* (1915), deployed advanced narrative storytelling techniques, albeit in the service of a deeply problematic and racially prejudiced narrative. Nevertheless, Griffith's work exemplifies how film began to tackle complex and controversial subjects, embedding itself in the socio-political dialogues of the time. Similarly, Russian filmmaker Sergei Eisenstein (1898–1948) brought a new layer of intellectual rigor to film. His groundbreaking works, such as *Battleship Potemkin* (1925) and *October* (1928), are noted for their innovative use of montage to create psychological and ideological impact. Eisenstein's approach to film was not merely as a form of entertainment but as a method of intellectual and political engagement. His theoretical writings, including the seminal text *Film Form: Essays in Film Theory*, published in 1949, have influenced generations of filmmakers and scholars alike (Eisenstein 2014).

Following the pioneering contributions of early filmmakers such as D. W. Griffith and Sergei Eisenstein, the subsequent generation of directors and auteurs further expanded the lexicon of cinematic language and broadened its thematic scope. As the medium evolved, filmmakers hailing from a myriad of cultural and ideological backgrounds began to exert their influence, enriching cinema with a diversity of perspectives and artistic approaches. In the 1940s and 1950s, Italian Neorealism emerged as a prominent film movement, with directors like Roberto Rossellini (1906–1977) and Vittorio De Sica (1901–1974) exploring the realities of postwar Italy through a documentary-style aesthetic. Works such as *Rome, Open City* (1945), and *Bicycle Thieves* (1948) provided trenchant social commentaries, employing non-professional actors and shooting on location to capture the verisimilitude of human experience. Parallel to this, the French New Wave of the late 1950s and 1960s marked another transformative moment in the cinematic landscape. Directors like François Truffaut (1932–1984) and Jean-Luc Godard (1930–2022) rejected the conventions of mainstream cinema to offer a more introspective and often experimental approach to narrative. In films like *The 400 Blows* (1959) and *Breathless* (1960), traditional storytelling norms were disrupted, emphasizing individual subjectivity and innovative editing techniques (Marie 2008).

Across the Atlantic, American cinema of the 1970s experienced its own period of rejuvenation. Directors such as Martin Scorsese (1942–), Francis Ford Coppola (1939–), and Steven Spielberg (1946–) introduced complex characters and intricate storylines, typified by films like *Taxi Driver* (1976) and *The Godfather* (1972). This era, often termed the "New Hollywood," upended traditional studio systems and emphasized directorial vision, leading to a surge in the production of films that were both commercially successful and artistically nuanced (Schatz 2013). Moreover, the advent of digital technology in the late twentieth and early twenty-first centuries catalyzed another wave of innovation, enabling filmmakers like James Cameron (1954–) and Peter Jackson (1961–) to explore visual and narrative realms previously considered infeasible. Cameron's *Avatar* (2009) took advantage of advancements in motion capture and 3D technology to immerse audiences in a fantastical world, effectively blurring the line between physical and digital realities. Peter Jackson's *The Lord of the Rings* trilogy (2001–2003) similarly leveraged the potential of computer-generated imagery (CGI) and digital compositing to adapt J. R. R. Tolkien's expansive literary universe into a coherent and compelling cinematic experience.

The advent of digital cinema also facilitated the democratization of film production and distribution, thus amplifying voices that were traditionally underrepresented in mainstream cinematic discourse. Independent filmmakers have become empowered by digital cameras and editing software, making it less financially prohibitive to produce high-quality films. Directors such as Ava DuVernay (1972–) and Jordan Peele (1979–) have harnessed these technologies to offer narratives that confront societal issues related to race, gender, and identity, as evidenced in films like *Selma* (2014) and *Get Out* (2017). Simultaneously, international cinema has witnessed a resurgence with the newfound ease of global distribution. Filmmakers like Asghar Farhadi (1972–) in Iran, with films like *A Separation* (2011), and Bong Joon-ho (1969–) in South Korea, with *Parasite* (2019), have gained global recognition, thus reinforcing the assertion that cinema serves as a universal language capable of transcending geographic and cultural boundaries. The accessibility of international films has broadened the horizons of audiences and scholars alike, enriching the global film canon with diverse perspectives and storytelling techniques (Lee 2020a).

The contemporary cinematic landscape is marked by its multinational and multiplatform nature, an observation supported by existing research. Advancements in digital technologies have left an indelible imprint on this field, ushering in intricate visual effects and novel models of distribution, most notably, and streaming services. As the frontier of technological innovations expands, cinema remains a focal point of academic scrutiny. A recurrent theme in scholarly discourse posits that emergent technologies, such as extended realities, stand to fundamentally alter the paradigms of film production and consumption (Alizadehsalehi et al. 2020). The television series *The Mandalorian* serves as an instructive example of how these nascent technologies can be deployed for enhancing filmic narratives. The usage of LED walls in its production facilitated an integrative approach to computer-generated imagery, thereby enriching the viewer's immersive experience. It is worth noting, however, that the modality of audience interaction remains substantially unaltered in these instances. Yet, as technologies related to virtual and augmented reality mature, they may catalyze a transformation in the viewer's role from passive spectator to active participant within cinematic narratives. Conversely, these technological evolutions also portend the attenuation of traditional cinema-going practices. The emergence of streaming services and home-based entertainment solutions has siphoned off a sizable portion of the audience that once frequented movie theatres. This shift toward digital platforms, coupled with their financial efficiencies, has elicited a decline in traditional movie theatre patronage.

Moreover, the ascendancy of serialized content—epitomized by popular series such as Netflix's *Stranger Things* and *The Witcher*—is another pivotal development shaping the industry's future (Erickson 2019). The inclination toward binge-watching and a sustained narrative dispersed across episodes or seasons underlines a transformative trend in consumer behavior, one that will indubitably influence the evolving structure and scope of cinematic art and commerce. The acceleration of shifts within the cinematic landscape has been catalyzed not merely by technological evolution but also by exigent circumstances, such as the global pandemic, and the advent of new distribution platforms like Disney+. Intellectual properties like the Marvel Cinematic Universe and Star Wars epitomize this reconfiguration as they straddle both episodic and theatrical formats, making Disney+ a nexus for serialized narratives as well as exclusive film premieres. Concurrently, the burgeoning practice of multi-platform distribution has engendered changes in marketing and release strategies for films. Within this evolving paradigm, the potential role of emerging technologies such as virtual and augmented reality in reinvigorating the cinematic experience has been underscored (Rach and Scott 2020).

The transformative capacity of technological innovations in the domain of film production and distribution extends beyond mere shifts in the cinematic landscape to reverberate through the broader spheres of art and culture. The ingress of digital tools, visual effects, and interactive formats not only facilitates innovative storytelling paradigms but also engenders a more participatory and immersive engagement with visual narratives. These advancements echo Benjamin's concept of "aura" and value in art as functions of their reproducibility and mass dissemination. Contemporary iterations of film, steeped in technological mediations, introduce new dimensions to Benjamin's discourse, encapsulating not merely issues of reproduction but also interactivity and audience agency.

Emerging technologies, in augmenting the capabilities of film as a medium, stimulate broader cultural dialogues concerning the intersection of art, technology, and society. The participatory potential of interactive narratives, for example, incites reconsiderations of art as a collaborative endeavor between creator and audience, thereby diluting traditional demarcations between "active" and "passive" engagements with art. Similarly, the democratization of production tools provides a more pluralistic platform for cultural expression, thereby enriching the corpus of visual narratives with diverse perspectives and themes. To encapsulate, the technological evolution within the cinematic domain does not function as an isolated phenomenon but acts as a catalyst, challenging and augmenting broader conceptions of art and culture. The transformative effects on film thus serve as a microcosm, revealing the mutable and adaptive nature of art in an age increasingly mediated and shaped by technological innovations.

3.5 Machine as Muse: The Ascendance of Machine Art

The rise of computer-generated imagery and digital art in the 1960s revolutionized the art world, pushing the boundaries of traditional artistic practices and opening new avenues for creative exploration. Building upon the transformative advancements of printmaking and photography, artists embraced the potential of new technologies to redefine artistic expression. Notable pioneers such as Frieder Nake (born 1938), Georg Nees (1926–2016), and Vera Molnar (born 1924) played a crucial role in this transformation by exploring algorithmic and computational approaches to art. For instance, Georg Nees, an influential figure in the early digital art movement, devoted his artistic practice to the utilization of algorithms in generating visually captivating forms and patterns. In his work *Sine Curve Studies* (1969), Nees employed mathematical formulas to construct intricate and precise geometric shapes. The composition of the artwork centers around the mathematical concept of sine curves, which manifests as a series of curved lines that give rise to sinuous patterns and variations. These curves are created through algorithmic calculations done on the computer, with each line representing a segment of the sine function. The execution of these curves exhibits meticulous precision, resulting in smooth and seamless transitions between segments, evoking a sense of fluidity and grace. The curves' repetition and variation contribute to the overall dynamism of the composition, infusing it with a harmonious visual rhythm. Through his exploration of algorithms and mathematical calculations, Nees expanded the realm of artistic expression, demonstrating the potential of computational methods to create visually captivating and mathematically precise artworks.

Vera Molnar, an artist originally from Hungary, embraced the computational nature of art creation and utilized algorithms and mathematical formulas to produce geometric abstractions. Her artwork *Interruptions à recouvrements* (1969), seen in Fig. 3.7, serves as a prime example of her exploration of systematic structures and the fusion of mathematical concepts with artistic expression. As a pioneering figure in computer art, Molnar harnessed mathematical algorithms to craft intricate and visually captivating compositions. The composition revolves around a grid structure, where intersecting lines converge to form a network of geometric shapes. The artwork comprises a series of repetitive patterns, each defined by precise mathematical calculations and algorithmic rules. The lines intersect and overlap, engendering a sense of complexity and visual depth within the composition. The scale and format of Molnar's work can vary, as it can be produced as a print or displayed on a computer screen, showcasing her adaptability to different mediums and technologies.

The artistic potential of mathematical algorithms and computational processes is a central focus in the works of these artists. Through their innovative approach to computer-generated art, they defy conventional notions of artistic creation and embrace the opportunities presented by technology. These early pioneers played a crucial role in bridging the gap between art and technology, showcasing the creative power of computer-generated imagery. Their artworks not only challenge established artistic practices but also establish the groundwork for the rise of the digital art movement. By embracing algorithmic and computational approaches, they paved the way for future artists to explore the vast possibilities of technology in artistic creation (Paul 2023).

The introduction of software tools like Adobe Photoshop and Illustrator in the 1980s brought about a profound transformation in the realm of digital art, heralding a new era of artistic exploration. These powerful programs revolutionized the creative process by providing artists with unparalleled capabilities for image manipulation and creation. With the advent of these tools, artists gained access to a vast array of digital brushes, filters, and effects, empowering them to experiment with color, form,

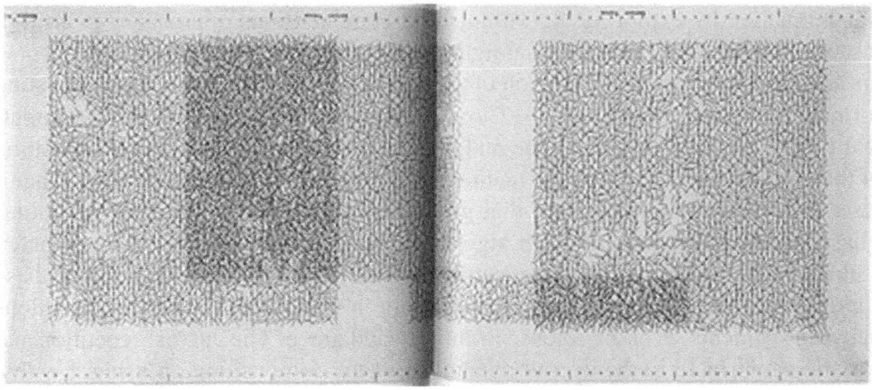

Fig. 3.7 Vera Molnar, Interruptions à recouvrements, 1969 (from Wikimedia Commons, licensed under CC0)

and composition in ways that were previously unimaginable. Adobe Photoshop, first launched in 1988, quickly emerged as an indispensable tool for digital artists with a remarkable new ability to manipulate and enhance images that opened up a world of creative possibilities. Artists could seamlessly blend multiple images, modify colors and textures, and apply diverse effects to produce visually striking compositions. With the aid of Photoshop, digital artists could freely experiment and iterate their ideas, pushing the boundaries of artistic expression to new frontiers (Sugiarto et al. 2021).

The advent of Adobe Illustrator in 1987 brought about a revolution in the realm of digital art, particularly in the creation of vector-based graphics. This groundbreaking software empowered artists to produce precise and scalable illustrations, granting them the flexibility to work across various mediums and sizes. With Illustrator, artists could now craft intricate and detailed digital artwork that maintained its image resolution regardless of scale. This newfound capability enabled artists to explore novel forms of visual expression, blurring the boundaries between traditional art forms and digital creations (Wang 2021). Consequently, the introduction of these software tools had a profound impact on the field of digital art with certain artists wholeheartedly embracing the digital medium, pushing the limits of creativity and challenging conventional art forms. These practitioners blended digital and traditional techniques, seamlessly incorporating elements from photography, painting, and illustration into their digital compositions. As a result, the distinction between traditional and digital art became increasingly blurred as artists harnessed the unique capabilities afforded by these software tools.

The advent of software such as Photoshop and Illustrator not only revolutionized the field of digital art but also democratized it. With these tools, artists were no longer reliant on expensive equipment or specialized training to create digital works. The accessibility and versatility of these software programs opened up the digital medium to a wider range of artists, allowing them to explore and express their creative visions. The impact of software tools, such as those developed by Adobe, on the field of digital art cannot be overstated—they transformed the creative process, empowering artists to manipulate images, venture into new realms of expression, and blur the boundaries between traditional and digital media. With the regular release of updated versions in the decades since their introduction, these tools continue to shape the landscape of digital art, providing artists with the means to push the boundaries of their creativity and produce visually stunning compositions that captivate audiences around the world (Zhang 2020).

During this era, a new art form known as new media art emerged, encompassing a diverse range of digital and multimedia art forms. Influential artists such as Nam June Paik (1932–2006) and Jenny Holzer (1950–) played pivotal roles in shaping this movement and exploring the potential of digital technology in their artistic practices. Paik, often hailed as the "father of video art," was a trailblazing figure in integrating television sets, video projections, and electronic devices into his installations. His artworks challenged conventional notions of art and questioned the role of technology in society. A striking example is his installation *TV Buddha* (1974), seen in Fig. 3.8, where a Buddha statue faces a television monitor displaying a live feed of the statue itself. This seminal installation merges technology and spirituality, exploring the intersection of Eastern philosophy and Western media culture. The artwork features a traditional Buddha statue in a serene, meditative pose juxtaposed with a futuristic-looking television set. This juxtaposition creates a captivating contrast between ancient and modern elements. Positioned in front of the statue, the television screen presents a real-time video feed of the Buddha itself, blurring the boundaries between the tangible and the mediated and establishes a reflective loop where the Buddha observes its own image on the screen, prompting contemplation on the nature of perception and representation (Lim 2019).

The artwork accentuates the contrast between the conventional and the contemporary through the deliberate choice of materials. The Buddha statue, crafted from stone or other traditional materials, stands in stark opposition to the television, representing the cutting-edge technology of its era. The intentional juxtaposition prompts viewers to contemplate the intersection of spirituality and the pervasive mass media culture that emerged in the twentieth century. Furthermore, light assumes a vital role within the artwork, as the radiant glow emitted by the television screen illuminates the Buddha, creating a striking visual contrast against the surrounding darkness. The interplay of light and darkness contributes to the overall sense of enigma and spiritual ambiance within the installation, evoking a profound contemplation of the sacred and the technological realms.

Building on this tradition, Jenny Holzer is celebrated for her thought-provoking utilization of text and LED displays, employing digital technology to convey social and political messages within public settings. In her site-specific installation, *7 World Trade Center* (2006), seen in Fig. 3.9, Holzer engages with the architectural space, stimulating contemplation on the significance of the location. The artwork features a continuous stream of scrolling LED text projected onto the facade of 7 World Trade Center, a prominent building situated in Lower Manhattan, New York. The text, rendered in large and bold letters, demands attention and encourages viewers to

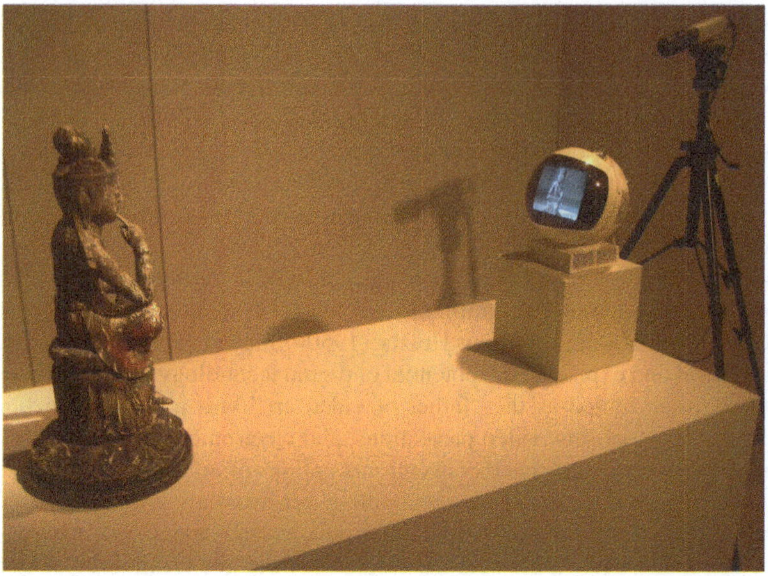

Fig. 3.8 Nam June Paik, TV Buddha, 1974 (from Wikimedia Commons, licensed under CC-BY 2.0)

actively engage with the conveyed message. The dynamic nature of the scrolling text creates an ever-changing visual experience, capturing the imagination and provoking introspection. The textual content within the installation holds profound significance, contributing to the meaning and resonance. The choice of words and phrases often tackles social, political, and cultural subjects. In the context of *7 World Trade Center*, the text delves into concepts related to memory, loss, resilience, and the process of rebuilding the World Trade Center site following the tragic events of September 11, 2001. The vivid and vibrant colors emitted by the LED lights serve as a striking contrast to the surrounding architectural elements, captivating viewers' attention and leaving a lasting visual impact. The movement of the scrolling text imparts a sense of urgency and dynamism to the overall composition, amplifying the potency of the message (Luger 2015). Through the interplay of scrolling text, radiant LED lights, and a site-specific approach, Holzer actively engages viewers in a dialogue surrounding social and cultural themes, while also acknowledging the unique context of the World Trade Center site. The artwork invites introspection, initiates meaningful conversations, and stands as a poignant testament to the enduring strength of the human spirit.

The emergence and proliferation of digital art and new media art were intricately linked to the rapid progress of computer technology during the latter half of the twentieth century. A new generation of artists embraced the potential presented by digital tools and techniques, propelling the boundaries of creative expression and embarking on uncharted paths of experimentation, interactivity, and multimedia integration. The utilization of computer-generated imagery, virtual reality, interactive

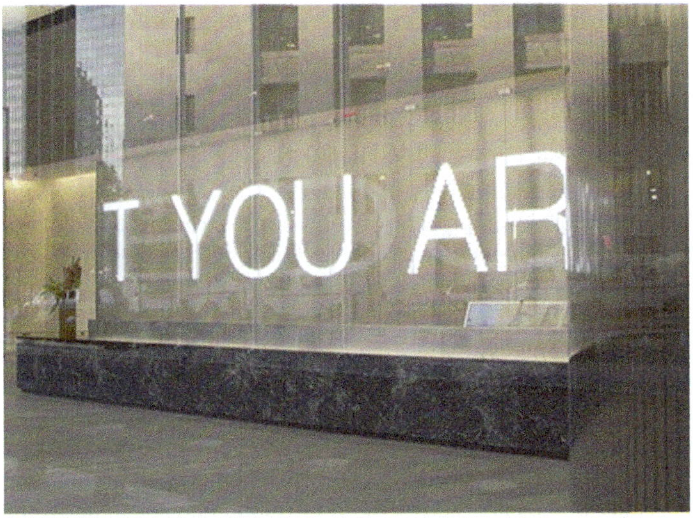

Fig. 3.9 Jenny Holzer, 7 World Trade Center. Photo taken November 2006 (from Wikimedia Commons, licensed under CC0)

installations, and Internet-based art became increasingly prevalent, broadening the horizons of artistic exploration and opening up new avenues for imaginative expression (Hope and Ryan 2014). This convergence of art and technology expanded the scope of artistic possibilities, enabling artists to engage with audiences in innovative ways and navigate unexplored realms of artistic creation. The marriage of art and digital technology ushered in a transformative period, redefining the boundaries of traditional art forms and ushering in a new era of artistic innovation and expression.

In parallel with the rapid advancement of digital art, there emerged a greater acknowledgment and embrace of this artistic form within the art world. Similar to photography, art institutions and practitioners were initially wary of art produced by digital means. However, museums and galleries, recognizing its significance, began to exhibit and acquire digital artworks alongside more conventional art forms. Institutions such as the Museum of Modern Art (MoMA) and the Whitney Museum of American Art played pivotal roles in showcasing and validating digital art (often referred to as "new media") through dedicated exhibitions and acquisitions. Their recognition and support provided digital artists with a platform to share their creations and contributed to the integration of digital art into the broader artistic discourse (Wands 2018). The newfound acceptance helped establish digital art as a legitimate and influential medium, fostering its continued growth and evolution in the ever-evolving landscape of contemporary art.

3.6 From Algorithms to Art: The Rise of Generative Art

The realm of art has witnessed a profound transformation with the advent of generative art, a form of artistic expression that harnesses the power of algorithms and computational processes to create dynamic and ever-evolving artworks. While receiving widespread attention only in 2022, the rise of generative AI can be traced alongside the parallel developments of digital art and the rise of new technologies since the 1950s. With generative approaches to artmaking, creatives and artists engage with algorithms as their creative partners, co-creating works that are in constant flux, responding to inputs, rules, and variables set by the artist or the computational system itself. This symbiotic relationship between artist and algorithm opens up new horizons of artistic exploration, pushing the boundaries of what is possible and redefining the very essence of creativity in the digital age.

The history of AI in art can be characterized by distinct themes and periods, each marked by significant developments and examples that have reshaped the artistic landscape. In the early stages of AI development, spanning from the 1950s to the 1990s, remarkable progress was achieved in the field of generative models. Two prominent models, namely Hidden Markov Models (HMMs) and Gaussian Mixture Models (GMMs), emerged during the 1950s, primarily utilized for generating sequential data. These models paved the way for generating data that followed specific sequences, enabling applications such as speech recognition and handwriting synthesis (Patel and Rao 2010). Further building on these advancements, the concept of N-gram language modeling gained significant traction in the 1960s and 1970s. N-grams refer to sequences of words or characters that frequently appear together in a given language. Language models based on N-grams opened the doors to sentence generation, enabling AI systems to produce coherent and grammatically correct sentences based on learned patterns and probabilities.

The advent of Long Short-Term Memory (LSTM) neural network architectures in 1997 marked a seminal moment in the domain of machine learning, particularly concerning sequence prediction tasks. The architecture, formulated by Hochreiter and Schmidhuber, was not designed for the translation of programming code, but rather for the efficient processing of sequential data with long-term dependencies, which is a central concern in natural language tasks such as machine translation of spoken or written languages. Given the fundamental challenge of modeling sequences in various applications, including but not limited to speech recognition and text generation, the utility of LSTMs extends broadly across multiple facets of artificial intelligence and computational linguistics (Gauthier 2014). With the incorporation of memory cells and gating mechanisms, LSTMs empowered models to learn complex dependencies over extended periods, leading to more accurate and contextually meaningful sequence predictions. The advancement in neural network architecture revolutionized the capabilities of AI systems in understanding and generating sequential data. These developments in generative models represented crucial milestones in the progression of AI, expanding its capacity to generate sequential data, including sentences and event sequences. The ability to produce coherent and contextually appropriate text has found applications in diverse fields, such as natural language processing, text generation, and dialogue systems. These breakthroughs laid the groundwork for subsequent advancements in AI, leading to the emergence of more sophisticated generative models and approaches.

The impact of AI on the visual arts became particularly profound in the 2000s and 2010s, ushering in a new era of image generation and manipulation. A pivotal breakthrough came with the introduction of Generative Adversarial Networks (GANs) in 2014, revolutionizing the field by enabling the creation of realistic and high-quality images through the interplay between a generator and a discriminator. The conceptual framework of GANs, introduced by Ian Goodfellow and colleagues in 2014, has engendered a profound transformation in the domain of computer vision and, by extension, in the visual arts (Goodfellow et al. 2014). The architecture comprises two neural network components: the generator and the discriminator, each of which serves a unique function within the adversarial framework. The operational dynamics between these two neural networks invoke an analogy to a forger and an art detective: the generator aims to produce counterfeit imagery so compelling that the discriminator cannot distinguish it from genuine data samples.

The generator network receives a random noise vector as input and processes it through a series of layers to produce synthetic images. During the training phase, the objective of the generator is to incrementally improve its capacity to generate fake data by minimizing the error of being classified as false by the discriminator. The generator's learning process is driven by the feedback from the discriminator, which guides it to produce increasingly realistic images over time. This architecture has found substantial applications in the visual arts, enabling the creation of hyper-realistic images, data augmentation, style transfer, and even the generation of art that challenges traditional aesthetics. The complex interplay between the generator and the discriminator within GANs has extended the frontiers of what is feasible in image generation and manipulation, thereby redefining the technological and conceptual boundaries of artistic creation.

GANs have since become a powerful tool for artists and designers, empowering them to generate visually captivating and diverse images that challenge the boundaries of traditional artistic creation. One of the most iconic examples of GAN-generated artwork is *Portrait of Edmond de Belamy*, created in 2018 by the Paris-based art collective Obvious. This piece was realized using a GAN trained on a dataset comprising portraits from the fourteenth century to the twentieth century (Stephensen 2019). Not only does the artwork exemplify the capabilities of GANs to generate high-quality and realistic images, but it also raises complex questions about authorship, originality, and the commodification of art, as evinced by its sale for $432,500 at Christie's auction house in 2018. The ability of GANs to synthesize new visual content has opened up new possibilities for artistic expression and has sparked a reimagining of the creative process in the digital age. In the same year, a notable advancement emerged with the arrival of variational autoencoders (VAEs). VAEs introduced a groundbreaking capability for compressing and decompressing data, thereby facilitating the generation of top-tier images. By acquiring latent representations of data, VAEs empowered artists to delve into diverse visual styles and produce innovative compositions. Through the utilization of VAEs, artists could explore uncharted territories of visual expression and unlock the potential for creating unprecedented artworks.

The emergence of VAEs has further enriched the landscape of machine-aided artistic creation, introducing another layer of complexity and opportunity. VAEs and GANs, while both generative models, operate on distinct mathematical principles and mechanisms. GANs rely on a dualistic framework involving a generator and a discriminator engaged in a sort of competitive game, wherein the generator aims to produce synthetic data indistinguishable from real data, and the discriminator seeks to distinguish between the two. VAEs, conversely, are rooted in probabilistic graphical models and are fundamentally designed for data compression and reconstruction. They operate by mapping input data to a lower-dimensional latent space, from which the original data can be reconstructed with high fidelity. This compressive capability enables artists to explore abstracted, condensed forms of data that can be subsequently manipulated and expanded into novel compositions. Moreover, VAEs offer a more controlled generation process, as the model learns a smooth latent space that can be navigated seamlessly, thereby allowing for more deterministic and interpretable manipulations (Sadeghi et al. 2020).

For instance, artists have used VAEs to interpolate between different styles in a more controlled manner, providing the capability to produce artworks that are a fusion of multiple visual styles. One example of artwork generated via a VAE is the project *Latent Space Cartography* by artist Memo Akten (1975–) (https://www.memo.tv/), wherein the artist traverses the latent space of a VAE trained on landscape paintings, thus creating evocative, otherworldly scenes that are at once familiar and disquieting. The use of VAEs in such works highlights the technology's power to offer a nuanced approach to data representation and generation, thereby opening up myriad possibilities for artistic intervention and innovation. Expanding upon these advancements, the year 2016 witnessed the introduction of StackGAN, a pioneering system that employed two GANs to generate high-resolution images based on text descriptions. This significant breakthrough enabled a more precise and intricate synthesis of images from textual input, empowering artists to manifest their visions with heightened fidelity (Zhang et al. 2018). Subsequently, in 2017, StyleNet emerged as a noteworthy development, aiming to generate captivating captions for images and videos with diverse styles. The innovative approach added a fresh dimension to the creative process, granting artists the ability to explore the dynamic interplay between visual and textual elements within their artistic works (Gan et al. 2017). Through the implementation of StackGAN and StyleNet, artists gained unprecedented tools to bridge the gap between textual descriptions and visually captivating representations, expanding the possibilities for artistic expression in the realm of generative AI.

The progress of AI in the 2000s and 2010s has revolutionized the visual arts and music composition, presenting artists and musicians with a plethora of new tools and techniques for creative expression. The emergence of GANs, VAEs, and other AI algorithms has pushed the limits of artistic creation, empowering artists to delve into unexplored visual styles, produce realistic images, and compose original musical compositions. The integration of AI in these domains has not only disrupted traditional artistic practices but has also paved the way for thrilling innovations and daring experimentation. More recently, remarkable advancements have been made in generative tools for artistic creation, further expanding the frontiers of AI-driven creativity. One notable breakthrough is the introduction of Midjourney, an AI text-to-image model that employs style transfer to metamorphose images and videos into diverse artistic styles. Additionally, Stable Diffusion, another text-to-image model, has made significant strides by focusing on developing stable and controllable generative models for image synthesis. Moreover, the expansion of the DALL-E model with DALL-E 2 now allows for the generation of high-quality images based on textual prompts, exemplifying the potency of language-guided image creation. Most recently, Adobe has introduced Firefly, a generative AI tool that empowers artists to experiment with dynamic and interactive visual effects in real time. These advancements serve as a testament to the continuous progress in generative tools, offering artists novel avenues for creative exploration while further blurring the boundaries between human and AI-generated art (Epstein et al. 2023).

The integration of AI into art production has unlocked a realm of uncharted possibilities for artists, challenging traditional perspectives on creativity and expanding the horizons of artistic expression. Leading the way, numerous artists have embraced AI as a tool in their creative processes, harnessing its capabilities to generate artworks that are both innovative and thought-provoking. Among them the aforementioned, Memo Akten, a London-based Turkish, stands out for his exploration of the artistic potential of deep learning models. His work *Distributed Consciousness* (2021), available at https://www.memo.tv/works/distributed-consciousness/, delves into the cognition of the octopus, utilizing AI to simulate synthetic alien intelligence. This project pushes the boundaries of AI-generated art by posing profound inquiries about the essence of intelligence and the intricate relationship between humans and machines. By leveraging deep learning models, Akten seeks to inspire contemplation regarding the intricacy and diversity of consciousness, surpassing the confines of human comprehension (Deterding et al. 2017).

The integration of robotics into the realm of art has experienced a remarkable surge, as artists leverage these technological tools to stretch the limits of creative expression. A noteworthy figure in this field is Sougwen Chung, a multidisciplinary artist born in China, raised in Canada, and currently based in London. Chung has made significant contributions by developing a unique and collaborative approach to artmaking with her ensemble of robots, collectively known as D.O.U.G. (Drawing Operations Unit: Generation_1). Detailed information about this project can be found at https://sougwen.com/project/drawing-operations. Powered by AI and recurrent neural networks, these robots learn to draw in the artist's distinctive style. This innovative fusion of AI and robotics facilitates a dynamic and interactive exploration of the creative process. Through captivating performances and installations, Chung initiates a dialogue between humans and machines, blurring the boundaries of authorship and challenging conventional notions of creativity. The collaborations between Chung and her robots transcend traditional ideas of artistry and delve into the concept of embodied AI. In working alongside the robots, Chung examines the potential of collaborative partnership, emphasizing the symbiotic relationship that exists between humans and machines. These thought-provoking interactions raise profound questions about the role of technology in art, the boundaries of creativity, and the evolving definition of artistic collaboration.

While some artists delve into the realm of cutting-edge technologies, others find a captivating equilibrium between traditional media and the possibilities presented by artificial intelligence. Linda Dounia, an artist and curator from Senegal, born in 1994, embodies this harmonious fusion in her artistic practice. Dounia seamlessly combines GANs with traditional materials, resulting in a distinctive blend of AI-generated and analog art. By training AI on her own creations by digitizing traditional media like painting and drawing and then uploading to a training model, Dounia explores the capacity of AI art to convey spontaneity and meaning. Her NFT (non-fungible token) project titled *Dust is Hard to Breathe* (2022) can be accessed at https://nft.artsy.net/artwork/linda-dounia/dust-is-hard-to-breathe/6/. Through this project, Dounia presents the curated outcomes of a generative adversarial network, engaging with profound themes of identity, resistance, and the limitations of AI-generated art (Boucher 2023). The exploration of identity and place with reference to Senegal stands at the heart of her work as she contemplates the intersection of human expression and the inherent constraints of AI. Harnessing the potential of generative adversarial networks, Dounia crafts thought-provoking art pieces that challenge conventional notions of artistic authorship and expand the boundaries of creativity.

Other artists engage in profound investigations of algorithmic biases through their artistic endeavors. British media artist Jake Elwes (1993–) employs AI as a means to question the training methods of computer systems and the biases inherent within them. His video piece titled *Zizi—Queering the Dataset* (2019) (https://www.jakeelwes.com/project-zizi-2019.html) intervenes by inserting the faces of drag performers into existing facial recognition datasets, purposefully challenging the limited representation of marginalized identities. Elwes intentionally disrupts these datasets to expose the limitations and biases embedded within AI algorithms, shedding light on the potential consequences of such biases in real-world applications. Through his work, Elwes not only critiques but also raises awareness about the importance of diverse and inclusive datasets in the development of AI technologies. He emphasizes the necessity of fair and equitable representation. Elwes's artistic exploration of algorithmic biases transcends mere critique. His art serves as a vehicle to empower marginalized communities, giving voice to those who have historically been underrepresented or misrepresented. Through the confrontation of the limitations of AI systems, Elwes initiates a dialogue about how technology can be harnessed to support and uplift these communities, rather than perpetuate societal biases. His work serves as a call to action, advocating for greater transparency, accountability, and inclusivity in the development and deployment of AI technologies.

3.7 Generative Worldbuilding: Shaping Immersive Experiences in the Metaverse

As AI technologies advance, new tools and algorithms are emerging that move beyond traditional two-dimensional artworks, allowing for the creation of immersive and expansive virtual environments on an unprecedented scale. One notable area where generative AI has made significant contributions is in the realm of worldbuilding. Artists and developers now have the ability to utilize AI algorithms to generate vast landscapes, architectural structures, and natural elements that populate the metaverse. From entire bustling cities like Angkor Wat to whole worlds with diverse ecosystems being constructed by the Digital Europe Programme, and even a galaxy as in Entropia Universe with many planets with fantastical landscapes to investigate discussed in Chap. 2, creating such content would be time-consuming and cost-prohibitive without these new abilities. These AI-powered tools can simulate realistic terrains, vegetation, and even dynamic weather systems, enabling the construction of intricate and lifelike virtual worlds. This capability to create on such a vast scale, and with such intricacy, opens endless possibilities for artistic expression and narrative exploration within virtual environments.

Moving beyond mere physical immersion, generative AI can now enable interactive experiences within these virtual worlds. Aside from the impressive visuals, large language models (LLM), such as ChatGPT, Bard, Llama, and others, can dynamically respond to user interactions, allowing for real-time modifications and adaptations to the virtual world. This interactive aspect adds a new layer of engagement and participation, blurring the boundaries between the artist and the audience, as users can actively shape and influence the virtual environment through their interactions (Seo and Alberse 2023). As such, the metaverse, powered by generative AI, presents a unique platform for artists, developers, and users to collaborate and co-create virtual worlds that transcend traditional artistic limitations; it offers a space where imagination and technology converge, allowing for the exploration of boundless creativity and the realization of fantastical visions. As generative AI continues to advance, we can expect even more sophisticated tools and algorithms that will further enhance the possibilities for worldbuilding and artistic expression within these ever-expanding and interconnected virtual spaces (Ratican et al. 2023).

With the emergence of the expanding virtual worlds and the increasing demand for immersive content, the development of 3D modeling software has played a crucial role in meeting the needs of creators and developers. Originally developed in the 1960s, 3D modeling has evolved to become an essential tool for creating detailed and realistic digital representations of objects and environments. There are various types of 3D models, each suited for capturing specific information about a three-dimensional object, such as its design, size, appearance, texture, and even physical properties like weight, density, and gravity. Initially utilized in industries and industrial design, the use of 3D models expanded in the late 1990s to encompass video games and entertainment. The demand for 3D asset creation has only grown with the rise of the metaverse and the consumption of media, games, and entertainment in immersive environments (Cheng et al. 2022).

To create 3D assets, modelers often rely on specialized software such as Sketchfab, Blender, Maya, and Autodesk 3ds Max. These applications allow for the creation and viewing of 3D assets, providing a platform for artists and designers to bring their visions to life. Additionally, for enhanced optimization and interactivity, these assets can be imported into game engines like Unreal Engine 5 (Epic Games) and Unity (Unity Technologies) and placed within virtual environments. These game engines also feature marketplaces where developers can download and even sell 3D assets, offering opportunities for collaboration and resource sharing among creators (Korbel et al. 2022). However, despite the availability of pre-existing assets in these marketplaces, developers often face limitations due to the finite number of resources and the need to customize assets to fit their specific requirements. Each asset can take a traditionally trained 3D modeler hours if not days to complete; the laborious process limits stock. The demand for high-quality, editable, and reconfigurable assets continues to rise, particularly in industries focused on immersive content creation for the metaverse. An example of that can be seen in cinema. The use of game engines and 3D models in cinematics has emerged as a groundbreaking technology, pushing the boundaries of realism and transforming the film industry. As details in 3D models have advanced to the point of becoming indistinguishable from real-life, industries such as film have witnessed a significant rise in their adoption, leading to remarkable advancements in experimental filmmaking, including VR cinematics. Notably, traditional game development software like Unreal Engine 5 has become an indispensable tool for visual effects artists and filmmakers, allowing them to create immersive and lifelike virtual worlds in real time.

A compelling illustration of the impact of this technology is exemplified by the American space western television series, *The Mandalorian* (Disney+, 2019–). The series served as a pioneer in popularizing the use of game-engine technology, providing filmmakers with unprecedented capabilities to construct virtual sets that dynamically adapt to changing needs and camera positions while accurately representing lighting information. As opposed to green-screen filming, which replaces everything in postproduction behind the actors, LED screens actually project what is desired, such as a distant local with no need for costly postproduction edits. This innovative approach significantly streamlined postprocessing efforts as a substantial portion of the visual effects were captured in-camera and on set. Furthermore, actors have lauded the incorporation of these technologies, as they can now experience and interact with the virtual environments they can see around them, marking a departure from the traditional practice of acting against green screens (Dent 2021). In other words, actors can react to what is actually in their space instead of imagining or reacting to abstract props. As the film industry increasingly relies on these advancements, the demand for 3D models and virtual sets continues to grow. Moreover, with the growing prevalence of VR filmmaking, the impact of these technologies has the potential to revolutionize conventional directing practices, presenting a paradigm shift in the industry. The integration of game engines and 3D models has not only elevated the visual quality of films but has also opened up new avenues for storytelling and creative expression, ultimately reshaping the filmmaking process and providing directors with unprecedented control over virtual environments.

The adoption of advanced technologies in the realm of filmmaking has indeed engendered a wide spectrum of responses from industry professionals. On one end lie the traditionalists who maintain an unyielding allegiance to analog methods and practices, perceiving the incursion of digital technologies as an affront to the sanctity of their craft. Filmmakers such as Quentin Tarantino (1963–) and Christopher Nolan (1970–) have been vociferous in their defense of celluloid, characterizing it as an irreplaceable medium that offers an unparalleled richness of texture and authenticity. Nolan, for example, has likened the use of film to the alchemy of time and innovation, tradition and techne, a sentiment that captures the apprehension many "old school" filmmakers feel toward an unchecked embrace of new technologies.

However, a contrasting narrative unfolds when one considers filmmakers like Scorsese, who have managed to straddle the technological divide with finesse. Scorsese, although a stalwart in the industry with a career spanning several decades, has been amenable to employing digital innovations to serve his storytelling objectives. An exemplary case is the film *The Irishman* (2019), where advanced visual effects technology was utilized to de-age actors, thereby enabling a seamless narrative that traversed multiple timelines. Such application of technology does not supplant traditional filmmaking techniques but augments them, expanding the scope and capabilities of what can be narratively achieved. The utilization of de-aging technology in *The Irishman* substantiates the argument that new technologies can serve as tools of artistic expression rather than mere gimmicks or crutches. Despite the remarkable advances enabled by these new advances, especially in game-engine technology, the widespread adoption of 3D modeling remains limited to specialists due to the intricacies involved in the development pipeline. In traditional 3D modeling, the process begins with the creation of simple geometry, often represented as polygons composed of vertices in three-dimensional space following the Cartesian coordinate system (X, Y, Z).

Generating surface geometry requires a minimum of three vertices, and modern game models can comprise tens of thousands of polygons, equivalent to a substantial number of triangles. It is important to note that these polygon counts are optimized for performance, as real-time applications necessitate the processing and rendering of thousands, if not millions, of polygons on screen at a smooth frame rate of 60 frames per second. The complexity of the modeling pipeline necessitates a range of specialized skills and techniques. For instance, re-topologizing meshes is a common practice employed to enhance animation and performance. The process refers to the restructuring or reorganizing of the topological attributes of a meshed surface in three-dimensional (3D) models. This operation is undertaken with a myriad of objectives in mind, including but not limited to: the optimization of mesh complexity, the enhancement of rendering efficiency, and the facilitation of more effective deformation or animation. Essentially, re-topologizing alters the way vertices, edges, and faces are connected in a mesh, ideally without altering its overall shape or the spatial relationships among its constituent elements. Rendering tools such as normal maps are utilized to add fine details to models, while postprocessing effects are achieved through the use of specialized shaders. The creation and optimization of 3D models require expertise in these areas, underscoring the need for specialized knowledge and training. Furthermore, the usability and optimization of these models within game engines and other real-time applications are essential considerations. Optimizing models involves reducing polygon counts, simplifying geometry, and implementing techniques to enhance performance without sacrificing visual quality. The efficiency of asset creation and its seamless integration within real-time applications are critical factors that impact the overall user experience and performance of interactive media.

In the following sections, we will delve into the two key variables of the asset creation pipeline: the intricate process of model creation and the optimization techniques employed to ensure the usability and effectiveness of these models within game engines and other real-time applications. Through an exploration of these aspects, we can gain a deeper understanding of the challenges and considerations involved in the creation and implementation of 3D models for interactive media. In recent years, there have been notable developments in the acquisition and creation of 3D models, with advancements in technologies such as 3D scanning and photogrammetry (Sharif et al. 2022). One prominent technology is 3D scanning, which has seen significant improvements in terms of quality, accessibility, and portability. Unlike earlier versions that were large and cumbersome, modern handheld scanners (e.g., Artec 3D) and smartphone applications utilizing Light Detection and Ranging (LiDAR) technology (e.g., Scaniverse and Polycam) have emerged. These scanners, along with their associated software applications, enable users to create 3D models of objects of various sizes and even scan entire areas. The versatility of 3D scanning extends beyond entertainment and finds applications in industries such as engineering and law enforcement.

Photogrammetry is another method related to model creation, involving the generation of 3D models from photographs or other data. While not a new process, photogrammetry has evolved from its origins in the creation of 2D information derived from photographs, sonar, and radar and has been utilized in various applications, including the production of topographic maps. The application of the technology has expanded into fields like entertainment, where it has been employed in films such as *The Matrix* (Wakowski and Wachowski 1999) and video games. Both 3D scanning and photogrammetry offer valuable means of producing 3D models and assets, providing artists and creators with options to capture real-world objects and environments for use in various interactive media.

As technology continues to advance, the integration of AI with photogrammetry has led to improvements in model generation with the ability to fill in missing details from photographs (Amaro et al. 2021). However, there has been a notable shift in the role of AI in the art world, transitioning from a tool for assisting in art creation to a method for generating art itself. AI art generators like DALL-E 2 have garnered attention for their ability to produce intriguing and imaginative artworks (OpenAI 2022). Although early iterations of these models may have exhibited apparent errors to human observers, each subsequent iteration has demonstrated improvement, with newer models even capable of generating photorealistic art. This progression in quality and sophistication is evident in the evolution of these algorithms. Another significant aspect of utilizing AI in art is its capacity to improve and learn from previous versions. Artists are now leveraging these generated images as sources of inspiration and starting points for conceptualizing ideas and designs, leading many to speculate that it may not be long before AI assumes a substantial role in concept design work. The ability of AI to continually refine its algorithms and build upon previous iterations holds tremendous potential for expanding creative possibilities and pushing the boundaries of artistic expression. Yet, despite these groundbreaking advancements, a constellation of ethical, philosophical, and aesthetic criticisms has arisen, many emanating from practitioners within the arts themselves.

A principal contention pertains to the ontology of art: the essence of what constitutes an artwork and the ineffable qualities that distinguish artistic endeavors from other forms of human activity. Critics argue that the involvement of artificial intelligence threatens to efface the unique human imprimatur that is integral to the act of creation (Roose 2022). In essence, AI-generated artworks lack the intentionality, subjective experience, and conceptual depth that are often considered the hallmarks of human creativity. Furthermore, the utilitarian rationale often used to justify AI's integration into the arts—namely its efficiency and the expansion of creative boundaries—has been met with skepticism. Therefore, while AI does proffer a new frontier of possibilities in the arts, the critiques against its pervasive integration into creative processes serve as a reminder of the complex web of ethical, cultural, and aesthetic considerations that remain. These discussions illuminate the dialectic between technological innovation and artistic tradition, a dialogue that is likely to intensify as advancements in AI continue to permeate the creative world.

Nevertheless, the progression from 2D art generation to 3D content generation represents a natural evolution in AI capabilities. As a result, various AI systems have emerged that can take a prompt, often a simple text description, and transform it into a 3D model. These state-of-the-art (SOTA) models are trained to comprehend 3D space using image language models. However, one of the challenges in this area is the scarcity of diverse and comprehensive 3D datasets compared to their 2D counterparts. The limitations of existing 3D datasets hinder the ability to train AI effectively in generating 3D content (Ha and Song 2022). Similar to how photometry utilizes data to generate images, certain AI systems are now leveraging 2D pictures as inputs to produce 3D content. Notably, Nvidia's Instant NeRF and Kaedim are AI systems that utilize images as references to generate 3D models. Kaedim, specifically developed for 3D artists and those in need of concept-based 3D models, is still undergoing development and relies on human reviewers to ensure output quality. By analyzing concept design images from various angles, Kaedim generates a corresponding 3D model. This tool considers the technical requirements of the models, making it one of the few AI-based 3D model generation tools that incorporate such considerations, although it does require users to specify the desired model complexity.

As well, other notable systems, such as Nvidia's Neural Radiance Field (NeRF), employ a process rooted in inverse rendering. Instead of traditional rendering methods that focus on lighting, NeRF analyzes a collection of 2D images and reconstructs a 3D model by simulating how light interacts with objects in the real world. This approach allows for the creation of complete 3D scenes in a relatively short period. An exemplary application of NeRF technology can be found in the realm of virtual tourism. Through the analysis of multiple 2D images captured from various angles at a particular heritage site, NeRF can reconstruct a remarkably detailed 3D model of the site. This allows users to virtually navigate through the space, experiencing the intricate details and variations in lighting that one would encounter in the actual, physical locale. Such applications not only stand as a testament to the technological prowess of NeRF but also contribute to an ongoing discourse on the virtualization of space and experience in the digital age. These methods of 3D model creation are becoming increasingly user-friendly, with accessibility extending to free smartphone applications. As a result, the barriers associated with 3D model creation have largely been overcome, making the process more accessible and streamlined.

Another crucial aspect to consider in the proposed immersive development pipeline is the optimization of 3D models for real-time applications. While the aforementioned methods can generate highly detailed models, these models often consist of dense meshes that are unsuitable for efficient performance and animation (Touloupaki and Theodosiou 2017). The term "dense mesh" refers to a three-dimensional polygonal structure that consists of an exceedingly high number of vertices, edges, and faces. In computer graphics and computational geometry, these dense meshes are often utilized for the accurate representation of complex forms or textures. However, such high granularity comes at a cost: these meshes are computationally intensive to manipulate, render, or animate. Therefore, while offering a meticulous level of detail, dense meshes pose significant challenges for real-time applications, especially those necessitating quick rendering or dynamic adjustments, such as video games or interactive simulations. Thus, the challenges associated with dense meshes serve as a compelling point of inquiry within both the artistic and scientific communities, motivating the search for optimization techniques such as mesh decimation and re-topologizing to facilitate more efficient computational handling.

Even current processes in 3D modeling for entertainment, such as digital sculpture, face similar challenges. Popular software applications like ZBrush employ digital sculpting techniques with remarkable results. However, this process can yield dense meshes that require re-topologization. Retopology involves creating a lower-resolution and optimized version of the model that is suitable for applications like game engines. The high-resolution details are then reintroduced during rendering. This process often includes baking the higher-resolution details into color images, where the surface of the high-resolution mesh is represented in a texture. These textures, known as normal maps, have been a standard practice since the early 2000s, utilizing the red, green, and blue channels to control the XYZ information of how light interacts with the surface of the model (Tasdizen et al. 2003). The retopology process can be complex and time-consuming, prompting software developers to incorporate auto-retopology tools into popular 3D software applications. Among the leading innovators in graphics technology is Nvidia, a renowned graphics card manufacturer. Nvidia has been at the forefront of sponsoring and developing new graphics technologies for several decades, often showcasing their advancements at their annual technology conference. Given the significant role hardware plays in processing and rendering immersive realities and real-time applications, Nvidia also focuses on improving performance in these areas.

Two recent innovations from Nvidia, Nvidia NGX and Nvidia NeRF, are particularly relevant to this study. Both technologies introduce new and dynamic approaches to rendering. Nvidia NGX leverages Nvidia RTX video cards, combining hardware capabilities with AI and deep learning techniques to enhance performance and optimize graphic output. By loading an entire 3D scene from a previous design iteration, the NGX can create modern lighting and rendering solutions. This process can transform older graphics with lower resolutions, ensuring they appear sharp and clear on modern systems and resolutions. It is important to note that while this process enhances the textural graphics, it does not directly address the resolution of the models themselves. For higher-resolution models, proper retopology and optimized geometry are still necessary. To address the challenges of retopology, engineers at Epic Games working on Unreal Engine 5 have taken a new approach by introducing virtualized geometry systems that aim to eliminate the need for retopology altogether. The Nanite system, built on earlier research and iterations on processing and rendering high-resolution geometry in real time, enables users to directly import non-optimized high-resolution 3D models from sources like 3D scans or detailed digital sculpts. These files can be dynamically resized to ensure fast processing and rendering.

Nanite employs several innovative techniques to achieve this. One such technique is the real-time generation of multiple level of detail (LOD) copies of an object, each with varying polygonal complexity. As the viewer moves away from an object, the resolution decreases, and as the viewer gets closer, the complexity increases. Additionally, the virtualized geometry system automatically occludes polygons that are not visible from the perspective of the viewer, effectively rendering them non-processed or non-rendered. Adjusting the resolution of out-of-frame objects improves performance in real-time applications, as render engines often process geometry even when it is not within the field of view of the viewer. Nanite further enhances rendering by utilizing a system of virtualized textures that are automatically generated. This is possible through capabilities to offer virtualized geometry. Unlike traditional methods that require the pre-processing of graphical assets, Nanite dynamically renders only the polygons that are visible at a given moment in a scene. This technique significantly reduces the computational overhead, allowing for the real-time rendering of exceptionally complex and dense meshes without loss of performance or quality. Virtualized textures in the Nanite system operate under a similar paradigm. They are generated and loaded dynamically, ensuring that only the necessary texture data are in memory at any given time, thus optimizing resource allocation and performance. In essence, both virtualized geometry and virtualized textures are about a dynamic, on-the-fly approach to rendering, which is computationally efficient and allows for higher levels of detail.

The approach by Nvidia with NeRF systems focuses more on reconstructing 3D models from 2D data and simulating the behavior of light for photorealistic rendering. NeRF, in particular, leverages neural networks to synthesize new viewpoints from 2D images. While technologically advanced and capable of producing highly detailed models, Nvidia's systems tend to be computationally intense and may not necessarily be optimized for real-time applications in the same way that Nanite's virtualized systems are. Thus, the primary distinction resides in the contrasting methodologies for resource management and real-time rendering capabilities. Nanite emphasizes computational efficiency and dynamism, rendering only what is immediately necessary for a given scene. In contrast, Nvidia's systems prioritize detailed reconstructions that, although visually impressive, can demand a higher computational toll. Regardless, these optimizations are handled by the system itself, removing technical obstacles that could affect performance and allowing developers to focus on enhancing the viewer experience. With this system, and future developments that are likely to follow, many of the most time-consuming and technically challenging aspects of 3D asset creation are effectively eliminated.

The proliferation of AI tools and the shift toward generating 3D models provide developers and cultural institutions with an expanded range of options to choose from. These AI-powered tools, such as those offered by Nvidia, 3DFY, Sloyd, Luma AI, Masterpiece Studio, and Google DreamFusion, offer various approaches to generating high-quality 3D assets. For game developers and studios, the generation of 3D assets has traditionally been a complex and time-consuming process. However, with the assistance of generative AI, the production costs and timeframes associated with creating 3D assets can be significantly reduced. AI-generated 3D assets can be created rapidly and stored efficiently, addressing the challenges of cost and technical limitations.

It should be noted at this point that these tools are being used more commonly in arts and cultural institutions. In fact, the infiltration of advanced digital tools into the domain of fine arts and cultural institutions constitutes a noteworthy paradigm shift in the confluence of technology and art. Previously, the utilization of these intricate software systems was largely confined to the sectors of game development and specialized software engineering. However, the current landscape portrays a different scenario: the democratization of technology and its application in domains hitherto considered insular and traditionalist. Platforms like Unreal Engine, Unity, and Blender are now available without monetary barriers, and the plethora of robust training modules available free of charge have catalyzed a knowledge transfer from the tech-centric sectors to the art world.

Within this evolving context, museums and cultural institutions are actively participating in this technological renaissance. Some major museums have incorporated technology specialists into their staff to oversee projects involving VR, 3D modeling, and other interactive experiences. The strategic advantage of this approach is twofold: first, the internal staff develops a nuanced understanding of the artistic objectives and institutional ethos, and second, a culture of technological innovation is cultivated within the organization itself. Conversely, certain projects are indeed outsourced to specialized software studios, primarily when the technical complexity or required skill set exceeds the capabilities of the in-house staff. These collaborations often serve as an extension of the artistic vision of the institution, and they offer the specialized technical prowess needed to realize complex projects.

It is worth emphasizing that the accessibility of these technologies has prompted many museums and cultural institutions to invest in upskilling their existing staff.

Internal workshops, often conducted in collaboration with tech companies or facilitated through online courses, have become an emerging trend. These educational initiatives not only equip staff with the skills needed to operate these advanced tools but also open avenues for innovation and creativity within the institution itself. The AI tools mentioned above employ different techniques to generate 3D assets. For example, GET3D by Nvidia utilizes generative AI trained on 2D images to create high-fidelity 3D shapes with intricate details. Sloyd.ai focuses on real-time 3D asset creation specifically for gaming, providing optimized assets for integration into projects. Luma AI utilizes real-time imaging for reference, producing high-quality 3D assets suitable for augmented reality (AR) applications. These AI tools not only streamline the asset creation process but also offer functionalities tailored to specific industries and project requirements. For instance, Gepetto.ai focuses on building AI-centric gaming and filmmaking experiences, while Masterpiece Studio aims to provide a toolset for generating game-ready 3D assets. Each tool brings unique capabilities and features to the table, expanding the possibilities for developers and artists in creating immersive content.

Furthermore, the advancements in generative AI for 3D asset creation extend beyond traditional game development. These tools have applications in various industries, such as virtual reality (VR), augmented reality (AR), and metaverse development. They offer new avenues for content creation, enabling developers and cultural institutions to explore creative uses of 3D assets in immersive experiences. As the field of generative AI for 3D asset creation continues to evolve, we can anticipate even more innovative tools and techniques to emerge. With the increasing accessibility and effectiveness of AI-generated 3D models, developers and cultural institutions have a wide range of options at their disposal to enhance their projects and create compelling virtual experiences.

The advancements in AI generative content have witnessed unprecedented strides since 2022, opening up new possibilities for 3D asset generation (Grba 2022). Practicing artists are increasingly adopting AI technology to enhance their creative processes and explore new possibilities in their artwork. One notable example is the 3D artist Nikola Damjanov, who utilized AI algorithms and NVIDIA RTX technology in his recent venture into generative art (https://www.artstation.com/nikoladamjanov). The combination of AI and the specific hardware enables the artist to create a 3D-printed sculpture of a flower with intricate details and designs. Specifically, the NVIDIA Quadro RTX 6000 GPU, combined with AI rendering and denoising capabilities, facilitated the realization of complex aesthetic features inspired by microfossils and Radiolaria (a type of protozoa).

In the creative process, Damjanov established rules and relationships within the AI system to facilitate iterations and design changes. This allowed him to easily modify specific elements of the sculpture and observe the corresponding adjustments in the overall design. While challenges emerged during the project, such as testing the physical parts of the sculpture for structural integrity, Damjanov's adoption of AI technology significantly expedited his workflows. AI-powered denoising and rendering reduced the time spent on look development and provided faster feedback on the final render results. Damjanov also explored the use of game engines and implemented NVIDIA Deep Learning Super Sampling to enhance graphics performance. The increased speed and power of AI-enabled tools have given artists like Damjanov more time and freedom to explore their creativity and bring their artistic visions to life. By leveraging AI technology, artists can push the boundaries of their artwork and experiment with novel techniques that were previously inaccessible.

The accessibility of generative AI has expanded beyond individuals with extensive technological expertise or practicing artists, allowing a wider range of individuals to easily harness the creative potential of AI for their own creations. One remarkable example of the accessibility of AI-generated content to the general public is that of Australian instructional designer Andrew Wright. In his virtual exhibition, *The Other Us* (2022) (https://framevr.io/theotherus) each image was generated using natural language prompts in an AI generator, showcasing how AI-generated content is now within reach for anyone interested. Currently, generating the desired effect for each image in the exhibition requires trial and error with thousands of text prompt permutations. However, rapid progress is being made in this field, evident in the increasing number of freely available AI generators. These developments are empowering individuals to experiment and create AI-driven content without the need for specialized training or technical expertise.

The rise of AI-driven content and the increased accessibility of 3D design and development tools are driving significant disruption in the field. The combination of these technical achievements is paving the way for an alternative 3D asset creation pipeline, where developers can utilize commonplace technology like mobile phones for object scanning or AI systems for generating 3D content. This democratizes the 3D design and modeling field, eliminating the need for an experienced artist or designer and rendering current design and development pipelines obsolete. As a result, these advancements create more opportunities for users to create the 3D models they require, unleashing the creativity and potential of a wider audience. The democratization of 3D design and modeling brings new perspectives and ideas to the forefront, fostering innovation and expanding the boundaries of what can be achieved in the virtual realm.

This overview highlights the profound influence of technological advancements on the art landscape, encompassing the realms of both traditional and digital art. From the Renaissance to the contemporary era, each successive technological breakthrough has opened up new frontiers, reshaping the essence of artistic creation and redefining how visual culture is perceived. Throughout history, innovations like the printing press, photography, and digital art have revolutionized the way art is produced and disseminated, enabling artists to explore increasingly realistic, rapid, and even automated creative processes. As technology continues to evolve, the emphasis has shifted from the intrinsic value of physical art objects to the exploration of conceptual and ideational dimensions within visual expressions. Furthermore, the emergence of extended reality (XR) technologies adds yet another dimension to the artistic realm, blurring the boundaries between the virtual and physical worlds and offering immersive and interactive experiences to both artists and audiences.

Within this dynamic and ever-evolving landscape, artists have consistently demonstrated their propensity to challenge traditional paradigms and break free from established conventions. They fearlessly navigate unexplored territories, embracing technological advancements as vehicles for expanding the horizons of artistic expression. These visionary artists understand that art is not confined to a specific medium or conventional form, but rather a profound reflection of the human experience and an unyielding pursuit of discovery and ingenuity. In this context, generative AI emerges as the latest frontier in this ongoing artistic journey of exploration and redefinition. It embodies the spirit of artists like Manet, who sought to defy the norms of the artistic establishment through exhibitions such as the Salon des Refusés in 1863, challenging the very essence of what qualifies as art (Desnoyers 1863). By blurring the boundaries between human and machine-generated creativity, generative AI unleashes a realm of possibilities for artistic production, providing artists with new tools and techniques to delve into. It serves as a potent catalyst for innovation, empowering artists to stretch the limits of their imagination and bring forth works of art that were previously inconceivable.

Generative AI has not only transformed the process of artistic creation but has also revolutionized the way we experience and engage with art and culture. By harnessing the power of AI, artists can now create immersive and interactive experiences that transport audiences to new realms of imagination and emotion. This technology has opened up unprecedented possibilities for creating multisensory artworks that blur the boundaries between the physical and the digital, confronting our traditional understanding of artistic mediums. With the new technology, artists can create virtual and augmented reality experiences that invite viewers to actively participate in the artwork, becoming co-creators of the narrative and exploring it from different perspectives. Through the integration of AI algorithms, artworks can dynamically respond to the viewer's input, adapting and evolving in real time, creating a deeply personalized and immersive encounter. Moreover, generative AI has also facilitated the preservation and reinterpretation of cultural heritage. By leveraging AI algorithms to analyze and interpret vast amounts of historical data and artifacts, artists can recreate and breathe new life into lost or damaged artworks and historical sites. This technology enables us to engage with our cultural heritage in ways that were previously unimaginable, fostering a deeper connection and understanding of our shared history.

As we continue to push the boundaries of generative AI and its application in art and culture, it is important to ensure that ethical considerations and responsible practices are upheld. Artists, technologists, and society at large must collaborate to navigate the ethical implications and establish guidelines that prioritize the integrity of artistic expression and protect the rights of both creators and participants. In all, generative AI has unlocked new avenues for creating and experiencing art and culture in immersive and transformative ways and expanded the possibilities of artistic expression, empowered viewers to actively engage with artworks, and revitalized our cultural heritage.

References

Aali H (2021) French royal women during the restoration and July Monarchy: redefining women and power. Springer Nature

Alizadehsalehi S, Hadavi A, Huang JC (2020) From BIM to extended reality in AEC industry. Autom Constr 116:103254

Allan M (2008) Deserted histories: the Lumière brothers, the pyramids and early film form. Early Popul Vis Cult 6(2):159–170

Amaro NM, Pupo IP, Pérez Montesinos DF, Piñero Pérez PY (2021) Trends in photogrammetry and its integration with artificial intelligence. In: Conferencia Cientifica Internacional Uciencia, Oct 2021. Springer International Publishing, Cham, pp 417–431

Ansari T (2022) How AI transformed the art world in 2022, 30 Oct 2022. Analytics India Magazine (AIM). https://analyticsindiamag.com/how-ai-transformed-the-art-world-in-2022/

Benjamin W (2008) The work of art in the age of its technological reproducibility, and other writings on media. Harvard University Press

Boucher B (2023) 6 artists who were using artificial intelligence before ChatGPT. Art. https://www.artsy.net/article/artsy-editorial-6-artists-artificial-intelligence-chatgpt

Brittain B (2023) AI-generated art cannot receive copyrights, US court says. Reuters, 21 Aug 2023. https://www.reuters.com/legal/ai-generated-art-cannot-receive-copyrights-us-court-says-2023-08-21/

Burke K (2021) Print and the early modern playing card. Oxf Art J 44(2):183–205

Camba JD, Hartman N, Bertoline GR (2023) Computer-aided design, computer-aided engineering, and visualization. In: Springer handbook of automation. Springer International Publishing, Cham, pp 641–659

Carlson WB, Gorman ME (1990) Understanding invention as a cognitive process: the case of Thomas Edison and early motion pictures, 1888–91. Soc Stud Sci 20(3):387–430

Cheng R, Wu N, Chen S, Han B (2022) Reality check of metaverse: a first look at commercial social virtual reality platforms. In: 2022 IEEE conference on virtual reality and 3D user interfaces abstracts and workshops (VRW), Mar 2022. IEEE, pp 141–148

Chia L (2020) Printing and publishing in East Asia through circa 1600: an extremely brief survey. Mediaevalia 41(41):129–162

Coddington RS (2022) The camera is the eye of history. Milit Images 40(2(220)):56–59

Daguerre LJM (1839) An historical and descriptive account of the various processes of the daguerréotype and the diorama. Am Photogr Hist Soc

Dent S (2021) See how 'the mandalorian' used unreal engine for its real-time digital sets. Engadget. https://www.engadget.com/2020-02-21-mandalorian-ilm-stagecraft-real-time-digital-sets.html

Desnoyers F (1863) Salon des refusés: la peinture en 1863. Azur Dutil

Deterding S, Hook J, Fiebrink R, Gillies M, Gow J, Akten M, Compton K (2017) Mixed-initiative creative interfaces. In: Proceedings of the 2017 CHI conference extended abstracts on human factors in computing systems, May 2017, pp 628–635

Duncan-O'Neill E (2021) Painting Molière as an act of rebellion: Honoré Daumier, censored speech, and revivals of theatrical satire. Word Image 37(2):206–219

Du Preez A (2020) Approaching Aby Warburg and digital art history: thinking through images. In: The Routledge companion to digital humanities and art history. Routledge, pp 374–385

Eisenstein S (2014) Film form: essays in film theory. HMH

Epstein Z, Hertzmann A, Investigators of Human Creativity, Akten M, Farid H, Fjeld J et al (2023) Art and the science of generative AI. Science 380(6650):1110–1111

Erickson MP (2019) Old strategies in the new paradigm. Political economy of media industries: global transformations and challenges. Routledge, Oxfordshire

Füssel S (2020) Gutenberg and the impact of printing. Routledge

Gan C, Gan Z, He X, Gao J, Deng L (2017) StyleNet: generating attractive visual captions with styles. In: Proceedings of the IEEE conference on computer vision and pattern recognition, pp 3137–3146

Gauthier J (2014) Conditional generative adversarial nets for convolutional face generation. Class project for Stanford CS231N: convolutional neural networks for visual recognition. Wint Semes 2014(5):2
Goodfellow I, Pouget-Abadie J, Mirza M, Xu B, Warde-Farley D, Ozair S et al (2014) Generative adversarial nets. Adv Neural Inf Process Syst 27
Grba D (2022) Deep else: a critical framework for AI art. Digital 2(1):1–32
Ha H, Song S (2022) Semantic abstraction: open-world 3D scene understanding from 2D vision-language models. arXiv preprint arXiv:2207.11514
Halkyard S (2021) A lecture upon the shadow: William Henry Fox Talbot's the pencil of nature. PN Rev 47(4):0_2
Hazard AM (2023) Marketing the dead of Antietam: photographs of death as a cultural commodity. Civ War Hist 69(1):9–41
Hendrix SH (2019) Reformation by accident? Martin Luther's ninety-five theses of 1517. In: A sourcebook of early modern European history. Routledge, pp 153–155
Hirsch R, Erf G (2014) Exploring color photography: from film to pixels. Routledge
Hope C, Ryan JC (2014) Digital arts: an introduction to new media. Bloomsbury Publishing USA
Hutson J, Harper-Nichols M (2023) Generative AI and algorithmic art: disrupting the framing of meaning and rethinking the subject-object dilemma. Glob J Comput Sci Technol 23(D1):55–61
Hutson J, Lively J, Robertson B, Cotroneo P, Lang M (2023) Creative convergence: the AI renaissance in art and design. Springer Nature
Kontou T, Mills V, Nichols K (2022) Frederick Scott Archer, 'on the use of collodion in photography': the chemist, March 1851, pp 257–258. In: Victorian material culture. Routledge, pp 445–447
Korbel JJ, Siddiq UH, Zarnekow R (2022) Towards virtual 3D asset price prediction based on machine learning. J Theor Appl Electron Commer Res 17(3):924–948
Kremnitzer K (2022) Manet and photography: looking at Olympia. Hist Photogr 46(1):20–30
Lee N (2020a) The films of Bong Joon Ho. Rutgers University Press
Lee TC (2020b) Anatomy and academies of art II—a tale of two cities. J Anat 236(4):577–587
Lim TS (2019) Moving meditation: Paik Nam June's TV Buddha and its Zen Buddhist aesthetic meaning. Dao 18:91–107
Lipton L, Lipton L (2021) Fox Talbot's photography. In: The cinema in flux: the evolution of motion picture technology from the magic lantern to the digital era, pp 67–74
Low UW (2019) Then I saw: the influence of Albrecht Dürer's apocalypse as paratext. Relig Arts 23(4):341–359
Luger M (2015) Poetry as monument: Jenny Holzer and the memorial poems of 9/11. Mem Stud 8(2):183–196
Marie M (2008) The French new wave: an artistic school. Wiley
McCoy G (1972) Photographs and photography in the archives of American art. Arch Am Art J 12(3):1–18
Metzer G, Richardson E, Patashnik O, Giryes R, Cohen-Or D (2023) Latent-NeRF for shape-guided generation of 3D shapes and textures. In: Proceedings of the IEEE/CVF conference on computer vision and pattern recognition, pp 12663–12673
OpenAI (2022) Research. Retrieved 20 September 2023. https://openai.com/
Paigneau D (2019) Baudelaire, Delacroix, Nadar. Enjeux de la représentation de soi, de l'autre et du monde. Acta Iassyensia Comp 1(23):11–22
Patel I, Rao YS (2010) Speech recognition using hidden Markov model with MFCC-subband technique. In: 2010 international conference on recent trends in information, telecommunication and computing, Mar 2010. IEEE, pp 168–172
Paul C (2023) Digital art. Thames & Hudson
Pope N (2020) The reception of Kodachrome sheet film in American commercial photography. Technol Cult 61(1):1–41
Rach M, Scott R (2020) The use of virtual reality in marketing: exploring the need for technology and language adaptation to create high quality immersive experiences. In: Marketing and smart technologies: proceedings of ICMarkTech 2019. Springer Singapore, pp 67–77

Ratican J, Hutson J, Wright A (2023) A proposed meta-reality immersive development pipeline: generative AI models and extended reality (XR) content for the metaverse. J Intell Learn Syst Appl 15
Roose K (2022) An AI-generated picture won an art prize. Artists aren't happy. The New York Times 2, Sept 2022
Roughley M, Liu CYJ (2022) Digital 2D, 2.5D and 3D methods for adding photo-realistic textures to 3D facial depictions of people from the past. In: Biomedical visualisation, vol 11. Springer International Publishing, Cham, pp 245–280
Sabour WA (2021) Developing the waterless lithography printmaking process. Int Des J 11(1):393–402
Sadeghi M, Leglaive S, Alameda-Pineda X, Girin L, Horaud R (2020) Audio-visual speech enhancement using conditional variational auto-encoders. IEEE/ACM Trans Audio Speech Lang Process 28:1788–1800
Schatz T (2013) The new Hollywood. In: Movie blockbusters. Routledge, pp 15–44
Schwartz J, Ryan J (eds) (2021) Picturing place: photography and the geographical imagination. Routledge
Seo JH, Alberse J (2023) Connected immersion: interacting with dynamically generated creatures and narratives. In: Proceedings of the 15th conference on creativity and cognition, June 2023, pp 250–253
Sharif MM, Haas C, Walbridge S (2022) Using termination points and 3D visualization for dimensional control in prefabrication. Autom Constr 133:103998
Singh V (2018) The machine in the colony: technology, politics, and the typography of Devanagari in the early years of mechanization. Philol Encount 3(4):469–495
Snyder J, Allen NW (1975) Photography, vision, and representation. Crit Inq 2(1):143–169
Staub L (2019) Revisiting Renoir, Manet and Degas: impressionist figure paintings in contemporary Anglophone art fiction, vol 145. Narr Francke Attempto Verlag
Stephensen JL (2019) Towards a philosophy of post-creative practices?—Reading obvious' "portrait of Edmond de Belamy". In: Politics of the machine Beirut 2019. BCS learning & development, June 2019
Sturchio J (2020) Festschrift: experimenting with research: Kenneth Mees, Eastman Kodak and the challenges of diversification. Sci Mus Group J 13
Sugiarto E, Kurniawati DW, Febriani M, Fiyanto A, Imawati RA (2021) Computer-based art in folklore illustration: development of mixed media digital painting in education context. IOP Conf Ser Mater Sci Eng 1098(3):032017. IOP Publishing
Tasdizen T, Whitaker R, Burchard P, Osher S (2003) Geometric surface processing via normal maps. ACM Trans Graph (TOG) 22(4):1012–1033
Touloupaki E, Theodosiou T (2017) Performance simulation integrated in parametric 3D modeling as a method for early stage design optimization—a review. Energies 10(5):637
van der Wal J (2020) De fabrica humani corporis—fascia as the fabric of the body. In: Fascia, function, and medical applications. CRC Press, pp 3–18
Vigier D, Moncla L, Brenon A, McDonough K, Joliveau T (2020) Classification des entités nommées dans l'Encyclopédie ou dictionnaire raisonné des sciences des arts et des métiers par une société de gens de lettres (1751–1772). SHS Web Conf 78:11008. EDP Sciences
Wakowski L, Wachowski A (1999) The Matrix. Warner Brothers. (Film)
Wands B (2018) 25 years of curating digital art: 1993–2018. In: Electronic visualisation and the arts. BCS learning & development, July 2018
Wang R (2021) Computer-aided interaction of visual communication technology and art in new media scenes. Comput-Aided Des Appl 19(S3):75–84
Zhang N (2020) Application of computer graphics and image software in marine graphic design. J Coast Res 106(SI):600–604
Zhang H, Xu T, Li H, Zhang S, Wang X, Huang X, Metaxas DN (2018) StackGAN^{++}: realistic image synthesis with stacked generative adversarial networks. IEEE Trans Pattern Anal Mach Intell 41(8):1947–1962

Chapter 4
Digital Cultural Heritage Preservation

Abstract This chapter examines the intersection of digital preservation methods, such as 3D scanning and virtual reality simulations, with traditional conservation practices, highlighting how these tools not only protect but also extend the life and reach of cultural artifacts. It delves into the challenges and opportunities presented by the digital preservation of art and cultural heritage, including issues of authenticity, accessibility, and the ethical considerations of reproducing sacred and historically significant items. The chapter discusses the concept of "The Game of Culture," where game-based learning and interactive digital environments are utilized to educate and engage the public in cultural heritage, thereby fostering a deeper appreciation and understanding of diverse cultural histories. Furthermore, the narrative addresses the role of digital platforms in counteracting the effects of cultural erosion by providing alternative spaces for the exhibition and experience of endangered cultural assets. By highlighting examples from around the world, the chapter illustrates the global impact of digital technologies in cultural preservation efforts, suggesting that these digital endeavors not only preserve but also perpetuate cultural narratives in an ever-evolving digital landscape.

4.1 Globalization and the Constructed Worldview: Normalizing the Occident

Cultural heritage and its preservation are dependent on many hidden agendas. When considering the critical theme of digital cultural heritage preservation, Western perspectives and valuations of art and culture have profoundly shaped the preservation landscape. Of particular interest here are the factors contributing to the normalization of the Occidental perspective, perpetuating certain cultural narratives and valuations while marginalizing others. Additionally, the impact of gamification and game-based learning as potential disruptive tools is discussed, challenging and dismantling the hegemonic view of cultural heritage. Emphasizing embodied experiences as a means to understand and appreciate diverse cultures, this narrative highlights how they pave the way for a more inclusive and democratic approach to

J. Hutson, *Art and Culture in the Multiverse of Metaverses*, Springer Series on Cultural Computing, https://doi.org/10.1007/978-3-031-66320-8_4

digital preservation for the future of all peoples. Practical examples and strategies are presented to democratize technology, ensuring that cultural heritage preservation becomes accessible to a broader audience, fostering a more equitable representation of our shared human heritage.

The existence of museums and the field of art history play pivotal roles in shaping and upholding the fabric of the modern world. In the context of European modernity, Heidegger's evocative notion of the "age of the world picture" aptly describes a labyrinthine reality where pathways leading to exhibitions, fairs, theme parks, theatres, or museums seem to endlessly loop back into one another (Ivakhiv 2015). Building upon this, Timothy Mitchell's insights in "Orientalism and the Exhibitionary Order" (1989) shed light on the Parisian expositions of the nineteenth century, eliciting astonishment from non-European visitors, particularly those from Egypt. These exhibitions showcased the distinctively European inclination to transform the world into a representation—an exhibition. The driving force behind this fascination with the "organization of the view" (*intizam almanzar*, in Arabic) is rooted in perceiving the world as a system of objects, whose arrangement is believed to unveil deeper meanings or realities, such as Empire, Progress, or the Spirit of a People. This conceptual framework fostered a perception of tangible distinctions between objects as they exist and the meanings they symbolize—a phenomenon akin to what Heidegger termed "disposition" (Drabinski 1994). Such a worldview found harmony with an ethical perspective that viewed individuals as reflections of themselves within their works and deeds, while the approach often delineated individuals into separate "material" and "immaterial" aspects, where the latter was perceived as distinct from, or even transcendent of, its embodiment.

The phenomenon of delineating individuals into distinct "material" and "immaterial" aspects—wherein the immaterial transcends its physical embodiment—finds resonance in various instances across the annals of art and culture. One salient example arises from classical portraiture. In many traditional portraits, the material representation, that is, the physical features of the subject, is meticulously rendered, capturing the external likeness. However, juxtaposed against this material depiction is the endeavor to encapsulate the "immaterial" essence or spirit of the subject: their character, temperament, or soul. Leonardo da Vinci's *Mona Lisa*, for instance, captures not merely the physical appearance of the sitter but endeavors to represent an enigmatic spirit, a mysterious demeanor that transcends the mere brushstrokes on canvas (Kemp 2007).

Drawing upon the above exposition, Mitchell's reflections on the "organization of the view" elucidate a pattern, one where the art of displaying or arranging objects is underpinned by an ontological stratification. In this schema, objects—whether they be artworks, artifacts, or even human subjects—are not merely physical entities but are imbued with, or reflective of, deeper significations or realities. This bifurcation into the material and immaterial was not merely a passive observation but carried profound implications. As tangible entities were arrayed and exhibited, they became instruments through which intangible narratives—of empire, progress, or cultural spirit—were articulated and propagated. The "exhibitionary order," as

Mitchell termed it, thereby became an act of crafting perceptions, where objects served as conduits to relay profound narratives, aligning with the philosophical delineations of thinkers like Heidegger.

Recalling Hegel and his historical perspective on art, the unfolding of Spirit finds its legible traces or embodiment within art objects. This "museological" or "exhibitionary" order is a central theme to be explored when examining globalization and Mitchell contends that "orientalism" goes beyond being a mere portrayal of one culture (the Middle East/Asia) by another; it transcends the confines of colonial domination. Rather, it represents an integral part of the cognitive methods of order and truth—epistemological technologies, as described by Preziosi—that underpin the very essence of European modernity and forms an essential component of the cognitive organization of the modern European and Eurocentric world, giving rise to what can be termed an "exhibitionary order" (Preziosi 2009). Within such a world, a persistent desire arises to possess the "realities behind" the "representations" they encounter, a desire that may forever remain unfulfilled.

Globalization—the generic contemporary term referring to interconnected ideologies, policies, and practices supporting trans- and supranational flows of capital—itself constitutes a massively effective "organization of views," in this case the harnessing of diverse resources, interests, political, cultural, and social practices in the service of hegemonic economic order (Steger 2003). Global capital fabricates a virtual world order from the perspective of the relative economic value of practices and products of all kinds; an arena of investment opportunities (or foreclosures) where artistic practices of all kinds may be effectively staged as variable commodities with calculable exchange value. Among the palpable effects of this has been an increasingly explicit alignment of artistic production with current or potential market values, and the concomitant marginalization of practices not readily amenable to commodification.

An example of the increasingly overt alignment of artistic production with market values can be gleaned from the meteoric rise and commercialization of street art. Historically, street art began as an underground, often subversive movement, with artists like Jean-Michel Basquiat and Keith Haring leaving their marks on urban landscapes to make socio-political statements without the expectation of financial gains (Buchhart et al. 2022). However, with the globalization-driven "organization of views," street art underwent a transformative shift. A quintessential case in point is the enigmatic artist, Banksy. Initially recognized for guerrilla artworks on public walls, Banksy's pieces soon became eagerly sought after, fetching astronomical prices at auction houses. The very essence of street art, which was to remain accessible and free from commercial confines, became commodified. Artworks were physically extracted from their original locations to be sold to the highest bidder. The transformation of street art from a rebellious form of expression to a lucrative market commodity epitomizes the palpable effects of globalization on artistic practices and the stratification of artists based on their commercial appeal.

The disciplinary apprehension toward globalization manifests in multifaceted concerns, reflecting both the theoretical and practical implications for the domains of art criticism and history. One profound anxiety pertains to the potential erosion of distinct cultural art forms and practices. In an increasingly interconnected world, there is a lurking fear that localized artistic expressions, rooted in specific cultural, historical, or geographical contexts, might be overshadowed or assimilated by dominant global trends. Such an evolution could precipitate the loss of diverse artistic legacies, thereby impoverishing the tapestry of global art heritage (Makarova et al. 2019). Moreover, the establishment of a global standard in the discipline of art history evokes apprehensions about the homogenization of analytical frameworks and methodologies. A universally applied standard might prioritize certain artistic narratives, themes, or periods over others, consequently marginalizing art histories that do not conform to this predominant paradigm. Such a trajectory threatens to eschew the rich plurality of art historical traditions in favor of a monolithic narrative, potentially leading to a reductive understanding of global art and its multifarious trajectories (Elkins 2007). In essence, the challenges posed by globalization to the realm of art criticism and history underscore the tension between preserving the specificity of local art traditions and navigating the inexorable forces of global convergence. The overarching concern is that in the quest for a cohesive global narrative, the nuanced, diverse, and often discordant voices of local art histories might be silenced or subsumed.

However, amidst the extravagant claims about globalization, its genesis and evolution in early modern and enlightenment eras have been overshadowed, where attempts were made to articulate commensurability across diverse fields of knowledge and production. The development of art history as a European institution was intertwined with a vision of art as a global and pan-human phenomenon, even if the arts of Europe were perceived as the pinnacle of aesthetic evolution and cognitive sophistication—a notion not entirely separate from the early nineteenth-century Hegelian evolutionism of Spirit. Although achieving a truly "global" art history seemed like a distant multiculturalist goal, the current approach was fueled by the conviction that every culture, historical period, and geographic location deserved acknowledgment for their distinct and unique forms of "art." These creations were open to comparison and contrast with all other existing, lost, or potential artistic traditions or histories. Likewise, in museology, the belief prevailed that each collection of artworks or artifacts merely represented a fragment of an ideal totality, and no single institution could fully encompass it. Each collection was regarded as a glimpse into a future where global art would find its complete expression (Gorman 2013).

Globalization, a relatively recent phenomenon, can be traced back to the European expansion that began in the sixteenth century and has persisted with diverse implications ever since. This movement necessitated methods of translation and commensurability across various fields of knowledge and knowledge production, giving rise to numerous institutions such as museums, universities, and academic sciences. These institutions sought to address the challenge of finding common ground between different modes of knowledge within and across societies and cultural traditions. Art history, as an academic discipline emerging in the nineteenth century, embodied an attempt at universalizing cultural historiography, aligning with the teleological agendas of Western disciplinarity. An influential example of early globalization efforts was the 1851 Great Exhibition at London's Crystal Palace, Fig. 4.1, which

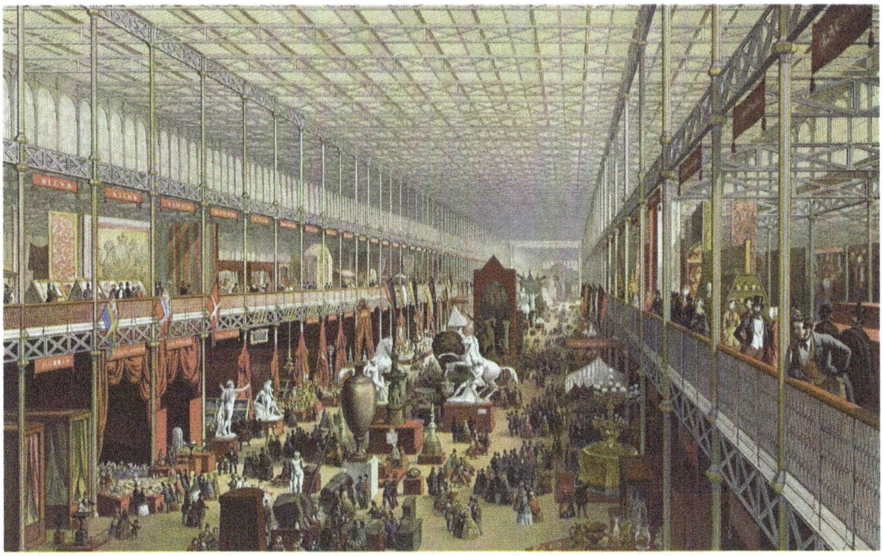

Fig. 4.1 J. McNeven, The interior of the Crystal Palace in London during the Great Exhibition of 1851 (from Wikimedia Commons, licensed under CC0)

showcased the arts and manufactures of diverse nations and contributed to the establishment of modern European institutions and professions in art history and museology (Auerbach and Hoffenberg 2008). Accompanying this exhibitionary order were "world histories of art," staged in academic and cultural institutions worldwide, along with extensively marketed introductions to the "art of the world." One illuminating case study can be found later in the Exposition Universelle (or Paris Exposition) of 1889.

4.2 Orientalism and the Exhibitionary Order: Understanding Cultural Perspectives

In "Orientalism and the Exhibitionary Order" (1989), Timothy Mitchell sheds light on how Western culture orchestrates the world's rationalization, comprehension, and control, which manifests in the design of European cities to accommodate expansive views, exemplified by the Haussmannization of Paris, the first modern city (Mitchell 2019). Notably, when Europeans ventured to Egypt, they encountered a lack of order and organization, leading to confusion. This historical context emphasizes the importance of considering cultural perspectives when approaching other societies. The relationship between the colonial order and the development of modern representation and knowledge has been thoroughly explored in the critique of Orientalism. The treatment builds on Edward Said's *Orientalism* (1978), which argues that Western scholars, writers, and artists have constructed a distorted and Orientalist representation of the East or the Orient. This representation perpetuates a false binary division between the West and the East, reinforcing the idea of Western superiority and Oriental inferiority. Said criticizes the Orientalist perspective for reducing diverse cultures and peoples of the East into a homogenous and exotic other and highlights how this Orientalist discourse has been instrumental in justifying and rationalizing European colonial domination and imperialism over the East.

The Saidian critique reveals that Western artistic and scholarly depictions of non-Western cultures are not mere ideological distortions but complex systems of imagery and expertise that shape the Orient as a political reality (Carrier 1995). This Orientalist reality hinges on three defining features: an understanding of the Orient as fixed by unchanging racial or cultural essences, the contrast of these essential characteristics with the West, such as passive-active, static-mobile, emotional-rational, and chaotic-ordered, and the marking of the Oriental Other with fundamental absences like reason or order. Through the employment of essentialism, otherness, and absence, the colonial world seeks to exert control, perpetuating and reinforcing these defining features.

Orientalism, as a part of a larger phenomenon, played a significant role in shaping the representation of the non-Western world in the nineteenth century. It was not confined to Oriental studies, literature, or colonial administrations alone; rather, it permeated all aspects of European efforts to organize and depict the world. This encompassed various domains, including museums, world exhibitions, architecture, education, tourism, fashion, and the commodification of everyday life. An exemplary event, the 1889 Exposition Universelle in Paris, seen in Fig. 4.2, attracted a staggering 32 million visitors, exemplifying the magnitude of these processes and showcasing French commercial and imperial power (Corbillé and Fantin 2020). The rise of Western hegemony, both economically and politically, was connected not just to Orientalist imagery, but also to the array of new mechanisms that conveyed the meaning and structure of the world during the imperial era.

World exhibitions, in particular, played a pivotal role in representing the non-Western world to Western audiences. Scholars have highlighted how this construction of otherness contributed to the shaping of national identity and imperial ambition (Hendry 2020). To delve into the relationship between representation, as a modern tool for meaning and order, and the construction of otherness vital to colonial endeavors, insights can be drawn from the accounts of non-Western visitors to nineteenth-century Europe. One such example is an Egyptian delegation attending the Eighth International Congress of Orientalists in Stockholm in 1889. Before reaching Sweden, they stopped in Paris to visit the Exposition Universelle, providing a detailed account of their encounter with the representation of their own otherness (Brendemoen 2020). Examining accounts from visitors in the Middle East, we can discern the distinctiveness of the modern representational order exemplified by world exhibitions. What Arab writers found in the West extended beyond mere exhibitions and representations; it was the world itself being transformed into an endless exhibition. This world-as-exhibition became a realm where the artificial, the model, and the plan were employed to generate an unprecedented effect of order and certainty. However,

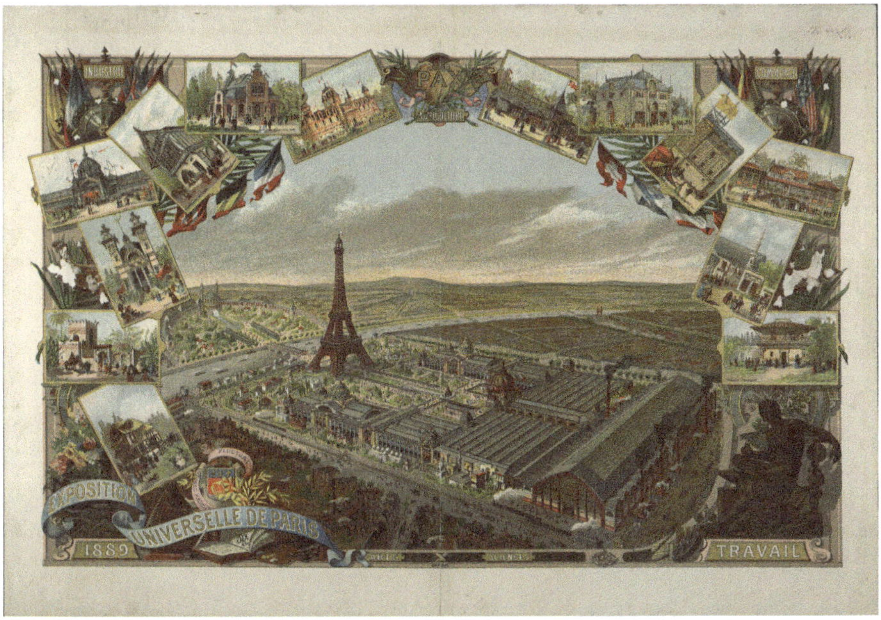

Fig. 4.2 Exposition Universelle de Paris 1889 (from Wikimedia Commons, licensed under CC0)

the true significance lies not in the artificiality of this exhibitionary order, but rather in the contrasting effect it created—a reality characterized, much like Orientalism's Orient, by essentialism, otherness, and absence.

The Egyptian delegation's experience at the Stockholm Orientalist conference and the subsequent visit to Paris highlights the pervasive Orientalist construction prevalent in European world exhibitions. The exhibit in Paris aimed to replicate medieval Cairo's appearance but fell into the trap of overemphasis on chaos and artificiality, further deepening the Orientalist perception. The Egyptians were dismayed by the inauthentic portrayal of their culture seen in Fig. 4.3, including the facade-like mosque transformed into a coffee house with dance performances and dervish rituals. Throughout the congress and exhibition, the non-European visitors found themselves constantly under scrutiny and treated as objects of curiosity. The European eagerness to observe and stare at them as if in a show displayed a recurring European concern with rendering the world for exhibition. This objectifying attitude persisted as Europe consolidated its colonial power, leaving non-European visitors subjected to degradation and display, contributing to the overall exhibitionary process that characterized European interactions with the non-Western world.

In addition to the curious attitude described in Arabic accounts, a corresponding objectness was intricately linked to the exhibitionary process of rendering things up as objects of fascination. Even Middle Eastern visitors themselves, Mitchell (2019) notes, were subjected to this objectification, as seen in the case of an Egyptian student mission in Paris, parodied in vaudeville for entertainment. The theatrical machinery

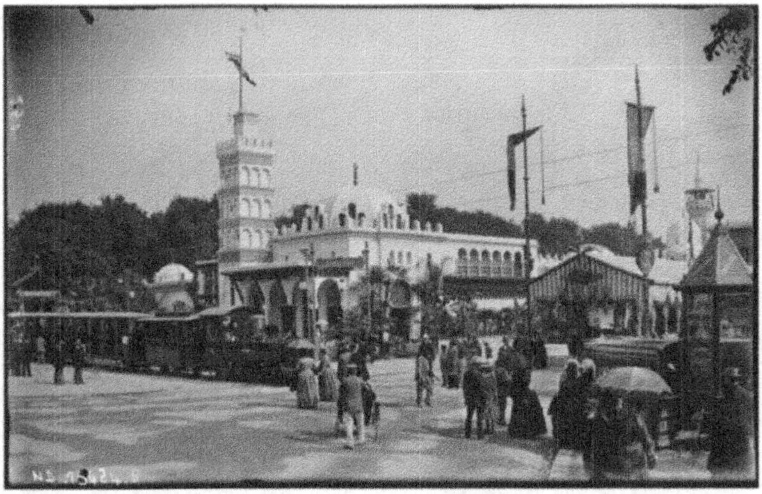

Fig. 4.3 Exposition Universelle de Paris, 1889. Pavillon de l'Algérie et le petit train (from Wikimedia Commons, licensed under CC0)

extended to Middle Eastern monarchs, like the Khedive of Egypt, who were incorporated into the European spectacle during their visits by being asked to stay in the displays of his home country. Moreover, the spectacle phenomenon, described with no Arabic equivalent, encompassed various presentations, including panoramas and dioramas, transforming the world into a picture for the dominant European gaze. European Orientalists, like Sylvestre de Sacy, sought to create similar object-worlds to allow European inspection of the Orient. In the late-nineteenth century, descriptions of journeys to Europe increasingly revolved around world exhibitions and Orientalist congresses, detailing the peculiar order and technique of these events (Nasirdinova 2022). This comprehensive method of organization can be best understood as representation, encapsulating the world-as-exhibition, where everything was showcased and investigated under the dominating European gaze, as demonstrated in the view from the Eiffel Tower of the exposition seen in Fig. 4.4.

Finally, the objectness experienced by Middle Eastern visitors went beyond mere visual arrangement and delved into the realm of representation. The careful organization of objects, whether in museums, theatres, public gardens, or zoos, aimed to evoke larger meanings like History, Empire, or Progress. Europeans lived in a world of spectacle and visual arrangement, preoccupied with *intizam almanzar*, the organization of the view. Everything seemed to be set up as models or pictures of something, arranged into a system of signification, declaring itself as mere objects, signifiers of further meanings (Godlewska 1995). The exhibition epitomized the peculiar character of the West, where individuals were continually made spectators in a world ordered to represent. This phenomenon presented the world to individuals as if it was on display, and visitors from non-European regions experienced this era of global display prominently in modern Europe. The spectacle and technique of

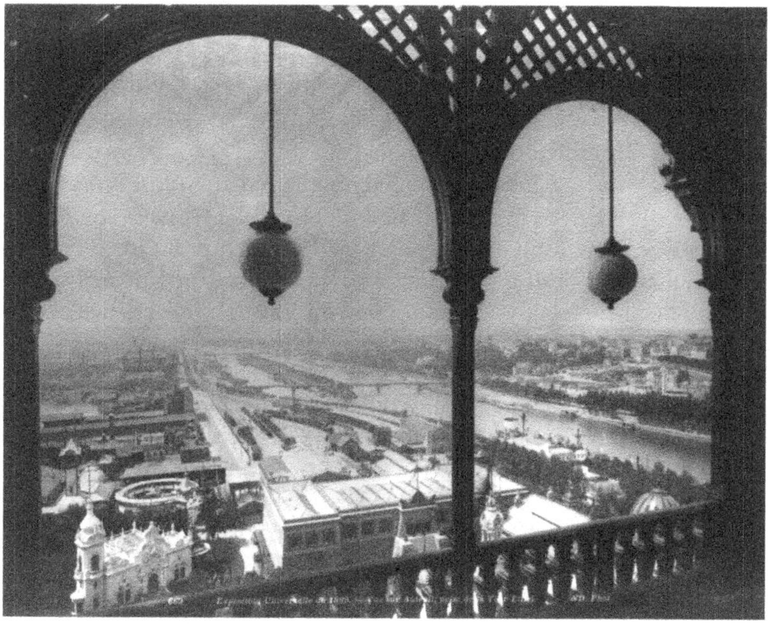

Fig. 4.4 View of Exposition Universelle from Eiffel Tower, Paris, 1889 (from Wikimedia Commons, licensed under CC0)

representation served to reinforce the political certainty of the imperial age, where England (given its control over vast swaths of the Orient, such as India) stood as the greatest Oriental Empire, knowing how to conquer and rule.

The world exhibition illustrated three aspects of certainty in the European representation of the Orient. First, the displays appeared remarkably realistic, seemingly corresponding perfectly to the external world. Second, despite their realism, the models remained distinguishable from the reality they represented, maintaining a deliberate difference in time and space. Third, within the exhibition, there was a distinction between the exhibits and the plans that organized them, reinforcing the separation between representation and reality. The boundaries between exhibitions and commercial structures in the city began to blur, particularly with the emergence and design of department stores. These department stores, with their grandiose displays, lavish layouts and deliberate presentation of goods, mirrored the ethos of exhibitions. Instead of merely housing merchandise, they crafted a visual and spatial narrative, immersing visitors in a meticulously curated environment.

Patrons were not only drawn to purchase but were also enticed to inhabit the world created within these stores. This desire for immersion was further evident in exhibitions that showcased foreign cultures, particularly the Orient. Visitors yearned to not only observe the Orient as a distant, exhibited entity but also to vicariously live within it, even if momentarily. This led to a paradoxical dynamic: While attendees wished to immerse themselves and be a part of the Oriental narrative, they concurrently perceived it as an object of marvel and exotic allure, maintaining a certain distance imbued with fascination. The exhibitionary spaces, in their quest to captivate audiences, thus walked a fine line between offering an immersive experience and perpetuating objectified representations.

Therefore, the world exhibitions embody a contradiction in European attitudes toward the Orient: the need to separate and represent the Orient as an object while simultaneously desiring to immerse oneself and experience it directly. The exhibitions cater to this dual desire, providing a profusion of exotic details while maintaining a clear distinction between visitors and exhibits. The experience of the Orient is shaped by the structure of the exhibition. The contradiction is also evident in how visitors like Gerard de Nerval sought to experience the real Orient within the labyrinthine streets of Cairo but found themselves unable to capture it until they moved to the outskirts of the actual city. This notion of the real versus the represented is not limited to Orientalism but underlies a broader division between the West and non-West, between a realm of representation and an external reality. The case of Orientalism reveals that this division is essential to the modern world's unique method of order and truth, shaped by the interplay of representations and reality, and it goes beyond colonial domination or cultural portrayal. The examples presented here illustrate the Western viewers' longing to interpret, comprehend, and claim ownership over the represented objects. However, the valuation process and decisions about what should be observed and preserved are often influenced by obscured factors.

4.3 Reevaluating Preservation Paradigms: Decoding Cultural Valuations

The cultural value assigned to heritage sites and monuments is not fixed but constantly evolving, influenced by various factors that can change over time (Gullino and Larcher 2013). A striking example of this phenomenon can be seen in the case of Victor Hugo's novel, *The Hunchback of Notre Dame*, and its profound impact on the perception and preservation of the iconic Notre Dame Cathedral in Paris. The Notre Dame Cathedral, seen in Fig. 4.5, was not always held in high regard by Parisians, especially after the Revolution when numerous treasures within the cathedral fell victim to destruction or looting, with only the great bells escaping being melted down. The Cathedral was rededicated to the cult of Reason and the cult of the Supreme Being and the interior repurposed as a storage facility for forage and food. The disregard for the structure was evident, but public opinion changed when Victor Hugo's novel was published in 1831 (Craven 2012).

The success of the publication of his renowned work, *Notre-Dame de Paris*, famously known in English as *The Hunchback of Notre Dame*, came out of the author's deep concern over the deteriorating condition of the cathedral, as well as

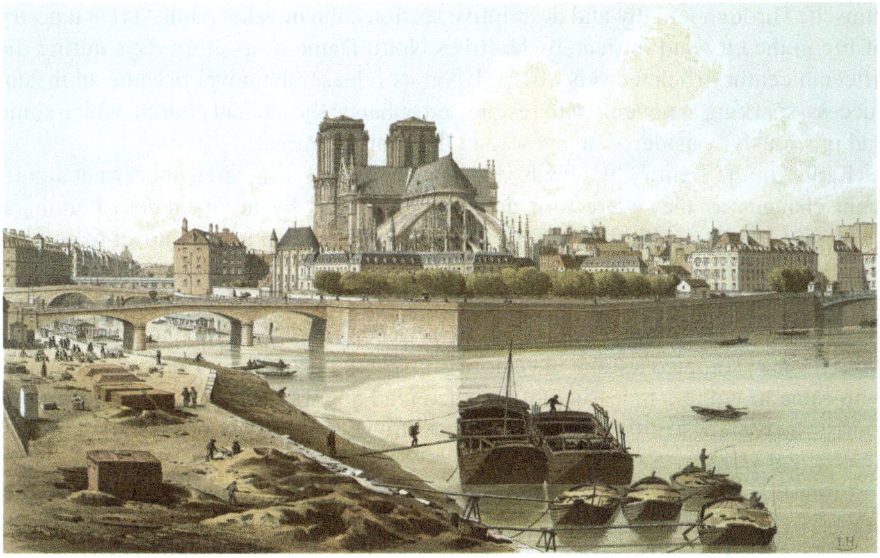

Fig. 4.5 Notre-Dame, l'Évêché et le cloître, 1830 (from Wikimedia Commons, licensed under CC0)

other neglected gothic monuments after the French revolution. With a determined goal to portray the church as a sacred masterpiece, denounce society's indifference toward its preservation, and inspire action, he set out to write a novel aligning with these objectives (Dubois 2023). Initially, his efforts faced challenges and distractions, but thankfully, much of his notes and the original manuscript for the novel were preserved, but it would not be until the second French revolution erupted in July 1830, a brief three-day upheaval that led to the overthrow of King Charles X, that Hugo's vision for his novel crystallized. Some speculate that the revolutionary spirit of the time ignited his determination, while others suggest mounting debts and impending legal action from his publisher served as additional motivators. Regardless of the reasons, Victor Hugo locked himself in his Paris apartment for a dedicated 4.5 months and diligently crafted the narrative that endures today (Spiqel 2006).

The prose of the novel eloquently captures the essence of Notre Dame, blending the sublime with the grotesque. The story unfolds in and around the declining church, and while it may be seen as a medieval tragedy with elements of a love story, it transcends mere melodrama. The principal characters include Quasimodo, a hideous and disfigured hunchback; Frollo, the Archdeacon of Notre Dame and the malevolent guardian of the protagonist; Esmerelda, a charming street dancer who captures the hearts of both Quasimodo and Frollo; and Phoebus, the captain of the king's archers and the lover of Esmerelda, who tragically falls victim to the schemes of the Archdeacon. However, the narrative is not limited to this captivating cast. Hugo also skillfully introduces a multitude of characters, offering us a vivid view of medieval Paris and its diverse inhabitants, from beggars to nobles, and even the King of France himself. Through lengthy and descriptive sections, the novelist paints a rich tapestry of life in the city and intricately describes Notre Dame in its glory days during the fifteenth century (Schneeweis 2016). Upon its release, the novel became an instant success, sparking a movement to restore and enhance the ancient church, which some had previously considered an eyesore in the heart of Paris.

Earlier in the century, the landscape of many gothic structures underwent significant changes as they were torn down and replaced by more modern buildings. Disturbed by this trend, the author purposefully sought to use his novel to emphasize the importance of preserving architecture as a valuable art form. As the novel gained popularity, it sparked a newfound romanticization of the cathedral, drawing attention to its historical and cultural importance. This heightened appreciation for Notre Dame prompted a surge in visitors from England and other European countries who came to witness the structure they had read about in person. As the number of tourists increased, so did the realization of Notre Dame's value as a symbol of French history and identity (Szwydky 2010).

From this surge in popular interest, the Commission on Historical Monuments organized a competition for the restoration of Notre Dame. Two talented young architects, Eugène Viollet-le-Duc (1814–1879) and Jean-Baptiste Lassus (1807–1857), emerged as the winners with their ambitious plan to restore the severely vandalized front facade of the cathedral seen in Fig. 4.6. The architectural vision of Viollet-le-Duc extended further, as he redesigned the spire of the church, as well. But the influence of Hugo extends far beyond the restoration of a church. The enduring popularity of *The Hunchback of Notre Dame* attests to its status as one of the most widely reproduced works of art in history. With over a dozen films retelling the captivating story and its alluring characters finding expression in paintings, engravings, sculptures, cartoons, songs, comic strips, graphic novels, stage productions, periodicals, and even podcasts, the novel has left an indelible mark on the artistic landscape (Martin 2011).

The popularity of the novel not only saved Notre Dame from the fate of other medieval structures demolished after the French Revolution, but the enduring fame of the beloved monument has once again played a role in preserving it. In 2019, the world watched live as a devastating fire engulfed the roof and spire of the iconic structure. The news of the burning Notre Dame spread rapidly, eliciting a global outpouring of grief and concern for the preservation of this cultural treasure, while also propelling the novel back to the top of worldwide best seller lists. The incident reignited collective awareness of the cathedral's significance, with people from different corners of the globe expressing their emotional attachment to this symbol of human achievement. Remarkably, the renewed cultural importance of Notre Dame in the aftermath of the fire was not merely confined to the people of France. Individuals and communities from various nations, who had experienced the cathedral through literature, history books, or personal visits, actively participated in discussions about its restoration and the allocation of resources for its reconstruction.

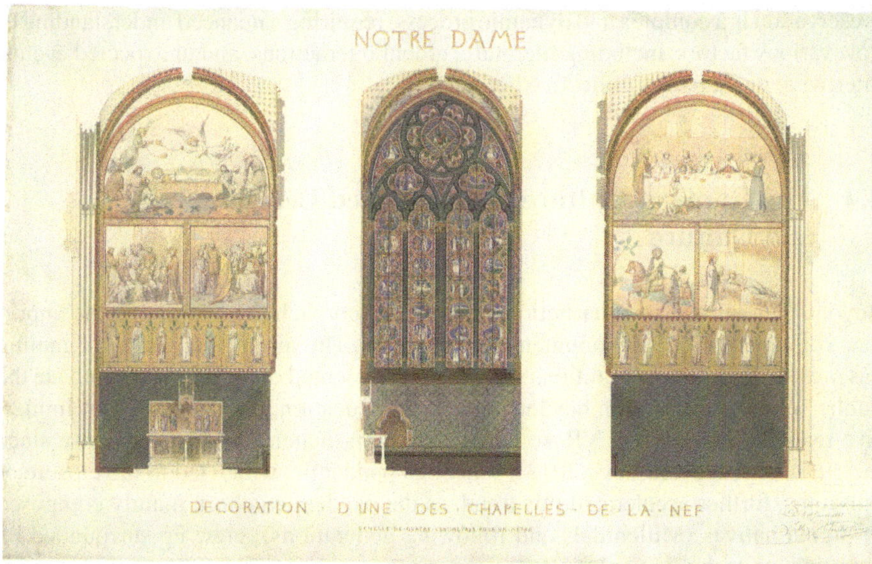

Fig. 4.6 Jean-Baptiste Lassus, plan for the renovation of a Chapel in the Nave of the Cathedral of Notre Dame, Paris, 1843 (from Wikimedia Commons, licensed under CC0)

The value placed on Notre Dame's historical and cultural importance led to the development of a full digital twin mapping of the cathedral (https://www.domusweb.it/en/architecture/2021/04/08/a-futuristic-simulation-of-notre-dame-to-power-its-real-world-resurrection.html). This cutting-edge technology involves creating a precise digital replica of the physical building, capturing its intricate details and architectural elements in a virtual environment. The digital twin enables a comprehensive analysis of the structure and its components, providing an invaluable resource for restoration and preservation efforts. The digital twin mapping was a unique undertaking due to the cathedral's fame—such as rigorous and expensive project, which is not something often performed on lesser-known churches or historical structures. By preserving its distinctive architectural heritage in a digital format, experts could conduct reverse engineering, spatio-temporal tracking of assets, and operational research, all contributing to the creation of an accurate reconstruction plan (Zhang et al. 2022).

This interaction between the narrative influence of Victor Hugo's literary work and the collective global lament over the near loss of Notre Dame illustrates the inextricable link between cultural valuations and preservation paradigms. The value attributed to heritage sites is not solely determined by the local inhabitants or the immediate society but can also be shaped by foreign peoples who develop a strong connection to these cultural landmarks, as well as popular culture recognition. This phenomenon raises important questions about how cultural valuations are established and whether preservation efforts should be reconsidered and expanded. The evolving perceptions and emotional attachments to heritage sites demonstrate that cultural preservation is a complex and dynamic process, requiring a nuanced understanding of how various factors, including literature, global interactions, and unexpected events, intertwine to shape our collective heritage.

4.4 The Game of Culture: Game-Based Learning in Art and Culture

Beyond the realm of popular fiction lies a novel approach to highlighting the importance of various cultural monuments and artworks. The surging popularity of gaming has presented cultural institutions with an unprecedented opportunity to captivate the public. While gamification has long enhanced education, the integration of immersive realities, particularly VR, to support learning materials has been in use since the 1990s (Biocca and Levy 2013). The 2020 pandemic, which led to the closure of museums, further accelerated this trend, as the pandemic cohort, mainly composed of digital natives (Millennials and following generations), grew up surrounded by cutting-edge technologies and never experienced a time without computers, smartphones, and 3D-gaming consoles. Consequently, the museum visitor experience has been profoundly influenced by technology, as people engage in socializing, shopping, gaming, completing homework, and more while strolling through galleries.

The digital transformation of museums builds on previous digitization efforts noted, such as with the Louvre and British Museum, and while the internet has broadened access to cultural heritage, museums can now take the next step and leverage emerging technology to offer personalized and interactive experiences to their visitors. An example can be found in the Israeli start up by Shani Ziv, Wandering, which utilizes a web-based platform for experiential learning with mobile devices, enabling visitors to choose specific items of interest and interact with related artifacts in the museum. One such application, Muse.run (https://muse.run/wandering/), provides interactive maps and digital games, allowing museum curators to design engaging experiences for diverse audiences. These innovations demonstrate how museums can combine technological approaches like smartphones, apps, interactivity, and gamification with their physical space to create a new era of museum-going. The essence of digital transformation lies not just in digitizing assets but in rethinking and redesigning the museum experience for personalized and sharable learning adventures.

The incorporation of games and gamification in the realm of arts and culture has gained considerable attention as a powerful tool to captivate museum visitors across diverse demographics, backgrounds, and cultures (Kim et al. 2018). Moreover, the concept of video games as an art form and the gamification of art history have become well-established ideas. Some early art games, like *Deus Ex* (Automata 1984), initially went unnoticed due to limited commercial success, and this is a common fate for many "art games." However, efforts have been made to redefine video games as art, as demonstrated by initiatives like the *Art History of Games Symposium at Georgie*

Tech in 2009 and 2010. Notably, the potential of video games to effectively engage audiences in art education has been extensively studied (Champion and Foka 2020).

While game-based instruction has been utilized before, the convergence of recreational gaming with easily accessible technologies has expedited the exploration and implementation of games for educational and cultural heritage purposes (Wideman et al. 2007). With the recent advancements in XR technology, educational gaming is primed to engage audiences in this new participatory, digital age. In the context of arts and culture, gamification holds tremendous potential for fostering visitor engagement. Educators and curators have grappled with maintaining interest and sustaining engagement, attempting various approaches, including motivational strategies, elaborate exhibition design, and marketing tactics. Unfortunately, these interventions often have only fleeting effects, while gameplay has demonstrated the ability to sustain attention and engagement over more extended periods. This emerging fusion of gaming and cultural experience has the potential to reshape how people interact with and learn from art and cultural institutions, fostering a deeper understanding and appreciation for our shared heritage.

Gamification is commonly understood as the incorporation of game elements into non-gaming contexts. For the purposes of this study, a more specific definition proposed by Kim et al. will be employed to provide a clearer understanding of educational gamification: "Gamification in learning and education is a set of activities and processes to solve problems related to learning and education by using or applying the characteristics of game elements" (Kim et al. 2017). This definition emphasizes that gamification is not a singular activity but rather a series of activities and processes. Unlike non-diegetic games, such as the Super Mario Bros series, which primarily aim to entertain, gamification must have specific educational objectives to address educational challenges effectively. Additionally, mere incorporation of game mechanics, like points or badges, is insufficient; gamification requires the inclusion of other game elements. The interactive approach to gamification of learning and game-based learning, despite sharing the common goal of promoting learning outcomes, diverges significantly in their design and implementation. Gamification of learning, as defined by Deterding et al. (2017), involves the application of game design elements in non-game settings. It does not entail the creation of a complete game but rather incorporates specific game mechanisms such as badges, points, leaderboards, and levels to enhance user engagement and learning. For instance, Shernoff et al. (2014) provided an example of a classroom environment where a system of points and badges was employed to motivate and reward student engagement and learning. This illustration demonstrates how gamification elements can be integrated into traditional learning contexts to foster motivation and engagement.

In contrast, game-based learning, as explained by Prensky (2003), involves the utilization of actual games, whether digital or physical, with an inherent educational purpose to facilitate learning. Unlike gamification, which borrows game elements, game-based learning entails the existence of a fully developed game intended for learning outcomes. For instance, Squire exemplified this approach by integrating the historical strategy game *Civilization III* (2001) into a high school history class, providing students with an immersive and interactive educational experience to grapple with complex historical phenomena and develop problem-solving skills (Squire 2006). Thus, while both approaches aim to harness the motivational appeal of games to enhance learning, they diverge in the level of integration of game mechanics into the learning experience. Gamification introduces game elements into traditional learning contexts, while game-based learning employs complete games to achieve educational goals. These distinctions hold significant implications for cultural heritage institutions as they deliberate on the most suitable approach aligning with their educational objectives and visitor engagement strategies.

Games and interactive experiences have emerged as potent tools for engaging and educating museum visitors, and they need not be intricate, expensive, or exclusively video game-based. In the domain of game design, it is crucial to acknowledge that while mechanics can be repurposed and adapted, originality and creativity play a pivotal role in delivering captivating experiences. By blending educational content, innovative gameplay mechanics, and a unique presentation, museums can craft games that offer a distinct and impactful experience for their visitors. Furthermore, Van Eck et al. (2015) contribute to the theoretical foundation of game-based learning by advocating for its role in creating a rich cognitive landscape for learners. His assertion finds support in three key examples: the game *Myst* (1993), aimed at cultivating problem-solving skills; the *Oregon Trail* (1971), which nurtures historical empathy; and the *Civilization* series (1991–2016), as previously noted, which fosters strategic thinking. These instances provide evidence that game-based learning, when effectively designed and implemented, serves as a potent pedagogical tool capable of enhancing the learning experience within cultural heritage contexts.

Similarly, the seminal work of Gee outlines the learning principles inherent in well-designed video games, positioning them as robust learning systems. Gee cites *Deus Ex* (2000), *Tomb Raider* (1996), and *The Elder Scrolls* series (1994–2014), highlighting how these games facilitate systemic thinking, the exploration of identities, and the development of a metacognitive understanding of the learning process. For instance, *Deus Ex* is a game that plunges the player into a multilayered narrative replete with political intrigue, ethical dilemmas, and choices that impact the game world. By navigating these challenges, players engage in systemic thinking, evaluating the potential consequences of their decisions in the broader context of the game's universe. This not only allows players to understand interconnected systems but also to consider the socio-political implications of their actions. The pedagogical merits of these examples align with the overarching goals of cultural heritage institutions, including fostering a deeper appreciation for culture and history (Gee 2011).

Expanding on this perspective, other research highlights the potential of gamification to foster inclusivity within museums. Museums and other cultural institutions have demonstrated the efficacy of game-based approaches in engaging visitors with successful examples like scavenger hunts at the American Museum of Natural History, interactive quizzes at the London Science Museum, and AR games at the British Museum (López-Martínez et al. 2020). These triumphs underscore the power of gamification in museums to attract and captivate diverse audiences. Through well-crafted hybrid game experiences, museums can offer interactive and educational opportunities that not only entertain but also deepen understanding and cultivate a lifelong appreciation for art, culture, and heritage.

The effectiveness of gamification in enhancing visitor engagement within cultural heritage institutions is evident from these diverse examples. The implementation of such gamified AR experiences at cultural heritage sites like the Tower of London demonstrates the potential for modern technology to breathe new life into history, making it more accessible, interactive, and memorable for visitors of all ages. It underscores the symbiotic relationship between preserving history and adapting to contemporary technological advancements. These instances, drawn from various sources, present a comprehensive outlook on the state of gamification in the field. As a result, arts and culture institutions should strive to further integrate game-based learning and gamification techniques into their visitor engagement strategies, considering the cognitive benefits and potential for increased inclusivity that these methods provide (Larson 2020). Embracing this initiative would not only make these institutions more appealing to a broader audience but also ensure a deeper appreciation and understanding of cultural heritage among their visitors.

Noting that 95% of cultural heritage institutions closed during the pandemic, Barbosa et al. (2021) found that despite the need to engage with audiences with new digital technologies, few actually integrated these. As the study found, the integration of technology into museums, especially in the form of dedicated mobile apps, represents an emerging trend to amplify visitor experience. While these applications can encompass a range of features, including maps, AR overlays, in-depth exhibit descriptions, and gamified components, their adoption is not yet ubiquitous. The prevalence of such mobile apps among science museums and centers, especially those affiliated with the ASTC or listed in the RedPOP publication, found that out of over 900 institutions under study (comprising 484 from the ASTC membership roster and 464 from the RedPOP guide), a mere 38 establishments had operational mobile apps tailored for smartphones. This translates to roughly 4% of the total entities. Such a statistic underscores that despite society's burgeoning embrace of digital technologies, science museums and centers remain somewhat tentative in their uptake of mobile apps as strategic tools to further their institutional mission.

The use of educational games and gamification instruction has garnered increased attention as a powerful tool to engage learners from diverse demographics, backgrounds, and cultures. Although game-based instruction is not a new concept, the widespread adoption of recreational gaming among a broader demographic, coupled with the ubiquitous availability of technologies, has accelerated the exploration and application of these games for educational purposes. With the recent advancements in XR technology, educational and cultural gaming is positioned to captivate students in the new participatory, digital age. In the realm of arts and culture, gamification holds immense potential for fostering learner engagement, addressing the persistent struggle educators' face in maintaining interest and attention in the classroom through various motivational strategies. In contrast, gameplay has demonstrated the ability to sustain attention and engagement for more extended periods (Kim et al. 2017).

The history of gamification in learning and education has been extensive, but budgetary and technological limitations have traditionally constrained the use of VR for educational purposes and gamification. In the early days of educational games like *Civilization*, 2D-level design and side-scrolling platformers were dominant, but a gradual shift toward greater immersion has since occurred. Initially, isometric design (illustration style that makes 2D objects appear three dimensional) was introduced for Massive Multiplayer Online Games (MMOGs) such as the *World of Warcraft* series (1994–), where players observed characters interacting with the environment from an elevated viewpoint. Subsequently, the first-person point of view gained popularity in First-Person Shooters (FPSs) like *Doom* (1996–) and became prevalent across various video game genres, reflecting the industry's endeavor to replicate the immersive qualities of VR (Wolf and Perron 2013).

Recently, XR has emerged as a commercially viable hardware option for consumers, and video games played a pivotal role in its development. The first attempt at promoting VR commercially can be traced back to second-generation video game consoles in 1993 when various video game studios developed software and applications for virtual reality head-mounted displays (HMDs). However, both SEGA (1993) and Nintendo (1995) faced commercial failure with their endeavors. It was not until 2013, with the introduction of the Oculus Rift, that the first commercially viable headset for VR emerged. While initially designed for the video game market, the Oculus Rift's capabilities also extended to support educational applications (Moorefield-Lang 2015). The latest generation of VR headsets, led by the Oculus Quest 2, has addressed the limitations outlined by earlier scholars, including user experience, accuracy, cost, and consumer availability. Unlike earlier educational VR games confined to Computer Science labs tethered to special desktop units, the latest all-in-one wireless headsets have expanded the distribution of educational applications through platforms like STEAM, Itch.Io, and Oculus App Lab.

As interest in exploring arts and culture through technology continues to grow, art history games present an exciting opportunity for broader public engagement. Four key areas are often highlighted in which such games can make significant contributions, including promoting an understanding of art classification, identifying artworks and their respective artists, providing insights into art creation and conservation, and conveying the contextual aspects of art creation and experience. Immersive realities, particularly VR, offer a powerful tool to transport students to various world cultural heritage locations and museums, enabling a deeper understanding of the context surrounding artworks. By incorporating interaction and sensory immersion, art history games can create thematic conceptual experiences that are enriched by contextual meaning.

Throughout their history, the arts have sought innovative technological methods to engage viewers with works. From printmaking and magic lantern projections to slide carousels and LCD projectors, technology has been utilized to disseminate reproductions of artworks. In the digital age, emerging technologies like VR have garnered significant attention in digital humanities scholarship due to their potential for preserving, representing, and disseminating cultural heritage (Bekele et al. 2018). However, unlike earlier passive information delivery systems, VR, coupled with gaming, offers a dynamic and interactive experience for audiences. Three key characteristics have been outlined that contribute to this engaging experience: establishing a contextual relationship between users, virtual content, and cultural context; enabling collaboration between users; and facilitating engagement with the cultural context in virtual environments. These features allow users to actively interact with the experience, collaborate with others, and gain a deeper understanding of the contextual relationship between art, its virtual representation, and the cultural environment within a virtual setting.

In addition to facilitating three-dimensional tours of monuments, VR has shown promise in aiding the understanding of abstract concepts, such as chronology, in the field of art history. One study involved students from the UK and Ukraine at different academic levels, from secondary to college, to investigate the effectiveness of VLEs in teaching historical chronology in diverse fields (Korallo 2010). The study highlighted the challenges in teaching the abstract nature of time to students at various levels. To address this pedagogical issue, the groups were taught the sequences of events within a virtual environment, while control groups were presented with the same events using traditional techniques such as texts, pictures, and PowerPoint slides. The results demonstrated that undergraduate students benefited the most when they could view three parallel timelines simultaneously in a continuous virtual environment, suggesting that using VR aids in comprehending historical events on a timeline and is more effective than conventional teaching methods. Another example of the pedagogical benefits of immersive technologies can be seen in the History of Western Architecture class (a compulsory course for architecture majors at Vellore Institute of Technology) where students were provided with the opportunity to study specific monuments in Google Earth VR, allowing them to experience the monuments in a human-scale, three-dimensional format rather than through two-dimensional images or digital projections (Ben Ghida 2020). This immersive approach has gained popularity in architecture departments worldwide, being adopted by institutions such as Utah State University, MIT, Queensland University of Technology, and more.

Furthermore, other immersive VR experiences have been developed such as for a Renaissance art history class at the University of Indiana, Bloomington. Four fresco cycles were created in the Unity game engine. Due to limitations in hardware and space, the VR experience was hosted in the Virtual Reality Lab on campus, where students explored the frescoes after covering the material in class. Teleportation was used to navigate between the different frescoes using VR headsets to minimize VR sickness. Teleportation, a commonly employed movement mechanism in virtual reality (VR), facilitates users' transition between different locations within the virtual environment without physically walking. By utilizing this method, users can swiftly "jump" to different points of interest, such as frescoes, in this context. This technique is particularly favored as it greatly reduces the risk of VR-induced motion sickness, a discomfort some users experience due to the disparity between visual motion cues and physical stillness. By integrating teleportation for navigation between the frescoes, developers aimed to enhance the user experience by minimizing potential discomfort. The VR experience also incorporated Smarthistory lectures triggered when approaching specific scenes, providing additional context and information. The production phase involved iterative playtesting to optimize the user experience (Brennan 2018).

The application of immersive technology and gamification in arts education has gained significant momentum. The virtual museum developed by researcher Huaman et al. (2019) exemplifies the benefits of gamification in the field, enhancing student engagement and learning outcomes. Within their virtual museum, interactive challenges and quests are designed around exhibited artworks, prompting students to engage more deeply with the content. As users navigate the museum, they encounter contextually relevant games or tasks related to specific art pieces, fostering active learning and encouraging critical thinking. Similarly, the online multiplayer educational game (otherwise known as a serious game) developed by Froschauer, Arends, Goldfarb, and Merkl, *ThIATRO* (2011) (accessible here http://vsem.ec.tuwien.ac.at/wp-content/thiatro.html), was created to address the challenges of teaching art history, including the need for extensive rote memorization and understanding cultural contexts. *ThIATRO* employs gamification to motivate students, fostering collaboration and active knowledge construction (Froschauer et al. 2011). The innovative, browser-based game immerses students in a 3D virtual environment to elucidate intricate art history concepts. As a "serious game," it intertwines entertainment with educational content, aiming to kindle a deeper enthusiasm for art history while fostering collaboration and active knowledge construction. Its design, rooted in 3D virtual principles, has been meticulously assessed with varied student demographics, underscoring its efficacy as an engaging learning platform.

The increasing popularity of software and applications, including game engines, has proven their utility in educational and cultural contexts. Several games have been developed to gamify art history, enhancing engagement, immersion, and content retention. Another game-based project by Casu, Spano, Sorrentino, and Scateni, *RiftArt* (2015) aimed to leverage cost-effective consumer hardware in developing an application for teaching art history (Casu et al. 2015). Serving as a VR tool enabling the configuration of virtual museum rooms with artworks, *RiftArt* also allows multimodal annotations; instructors can add additional multimedia content, such as audio or textual descriptions, providing a flexible and interactive learning environment. These virtual museum experiences offer unique advantages, allowing for comparisons between artworks housed in the same physical space while enabling visualizations of pieces from different locations. Another example can be found in the virtual museum created for the Universidad Nacional de San Agustin de Arequipa in Peru; their VR exemplifies gamification, allowing students to navigate through museum rooms, while earning points for interacting with the artworks. The positive survey data collected from students not only demonstrated improved learning outcomes but also highlighted the potential for greater interactivity and engagement by allowing students to create their own exhibitions. Meanwhile, *Art Thief*, a game created by the California Institute of the Arts (2017) (accessible here https://calarts-gamemakers.itch.io/art-thief), places players in the role of a young security guard named Olive, tasked with stopping an art thief in a museum and interacting with museum staff and visitors to solve puzzles. Although Art Thief demonstrates some gamification elements, it falls short in providing the full immersive experience found in other applications, limiting its overall engagement. To maximize the potential of gamification in arts education, the adoption of more immersive and interactive experiences, like those discussed earlier, holds great promise for enhanced learner engagement and deeper understanding of cultural heritage.

A clear distinction should be made between games and simulations. Many of the examples mentioned earlier are simulations, which differ from games in that they lack defined roles, goals, and agency. Notably, many art "games," such as *The Forbidden Palace: Beyond Space and Time* (2008), *Rome Reborn* (2018), and *Pompeii VR* (2021), fall under the category of simulations. In contrast, games like *Civilization* (1991–) provide players with specific roles (leader of a city), goals (advancing their people), and agency (their decisions impact the overall outcomes of the game). Simulations generally involve predetermined experiences intended to replicate or inform about specific situations, like learning how to drive a tank or experiencing a cinematic representation of the Pyramids of Giza. Both simulations and games serve as valuable learning experiences, depending on the desired educational objectives. However, while various examples demonstrate the potential of immersive VR experiences and game-based learning, there are limited instances where the two are effectively combined. The following example offers a model to bridge this gap, showcasing the innovative possibilities of integrating immersive VR with game-based learning strategies in the context of arts and culture.

4.5 Museum of the Lost: Embodiment and the Preservation Dilemma

Despite recent technological advances and the prevalence of a new participatory, digital culture that influences students at all educational levels, traditional teaching methods have remained largely unchanged. However, the concept of gamification in the arts, particularly experienced through VR, aligns well with this new participatory culture and holds the potential for transformative learning experiences. Studies have shown that motivating factors in gamified learning environments work in harmony with participatory learning, leading to improved motivation and academic achievement (Dastyar 2019). Gaming and gaming communities share similar characteristics, as both involve active participation and production of experiences through interactions with the game itself and other players.

To harness the interactive and immersive potential of games and VR for educational purposes, a multidisciplinary team consisting of game designers, coders, and art historians at Lindenwood University embarked on a novel project to develop a game that delves into the ethical complexities surrounding the protection of international cultural heritage and the repatriation of looted or stolen art. While using games to explore geopolitical issues is not a new concept—*Fantasy Geopolitics* (2016), *Geopolitics Boardgame* (2018), *Ultimatum* (1961), and *Spheres of Influence* (2016) are a few examples—the endeavor of aligning these considerations with the urgent need to safeguard cultural heritage had not been previously explored, despite mounting incidents of loss, such as the looting of Iraqi museums during the 2003 Iraq War, the destruction of Syrian monuments by ISIS from 2016 to 2017, and recent historical monument destruction in Ukraine by Russia from 2022.

Though international laws have been put in place to protect cultural heritage during wartime, debates surrounding preservation and repatriation persist. The complexity of these issues extends beyond cultural ownership and delves into the realm of ethics. While VR has been effectively used to teach ethics in medical and environmental studies, its application in addressing the ethical dimensions of cultural heritage protection remains largely unexplored. In response to these pressing issues, the serious game *The Museum of the Lost VR* (2022) was developed (available at https://bfulche.itch.io/museum-of-the-lost-vr) (Hutson and Fulcher 2023). The game offers historical overviews of cultural heritage protection and repatriation laws, while presenting the player with ethically ambiguous decisions that mirror the complexities faced by individual collectors, museums, governments, and nations in the present day. However, unlike similar-themed games such as *Art Thief*, the player is not the righteous protagonist looking to protect art from thieves, but instead is cast in the role of the thief themselves, immersed in an interactive environment, and forced to confront geopolitical complexities of theft and repatriation.

The game was developed using the Unity game engine, with a deliberate focus on casting the player in an ambiguous role. To effectively address the intricate issues surrounding artifacts from other nations and the ethical considerations of cultural ownership, the player is assigned the task of collecting specific works of art in three different sections of a museum functioning as a transit point for stolen or looted artifacts awaiting transportation elsewhere. Throughout the game, guards equipped with flashlights patrol all areas of the museum, posing a challenge for the player. The game mechanics incorporate stealth gameplay, requiring the player to navigate cautiously and avoid detection by the guards. In the event that a guard's flashlight illuminates the player's character/XR Rig, the player must respawn at the beginning of that particular level.

The main objective is to collect designated works of art, and a user interface (UI) element tracks the progress by removing each collected work from the repository—a crate used for shipping the artworks. Once all specified works are collected in each level, the player is presented with a quiz to test their knowledge on the history and legality of repatriation of cultural heritage. Three levels provide an historical overview of cultural heritage protection and repatriation laws, while presenting ethically ambiguous decisions to be made by the player, reflecting the complexities of issues facing individual collectors, museums, governments, and nations today. The Main Menu launches audio and an overview of the museum to introduce "What is a Museum?" The audio clip introduces the nature of global or encyclopedic museums and the practice of collecting objects from all geographical areas and chronological periods of human history. The checkered pasts of many objects in museums are introduced before the player spawns in the first level.

Level 1, entitled "The Hall of War: To the Victor Go the Spoils," delves into the historical context of looted artworks during the nineteenth to mid-twentieth centuries. It focuses on events such as the Napoleonic Wars (1803–1815), European colonization of Africa (1833–1905), and the World War II (1939–1945). In this level, the guards are not physically present; instead, their flashlight beams radiate across the museum floor, and the player must skillfully avoid these beams to progress. To navigate through the level, the player can use the crates scattered throughout the museum as cover from the guards' detection. The experience begins with a display of works looted by Napoleon during his European campaigns and his archaeological expedition to Egypt, which led to the establishment of the Napoleon Museum in the Louvre, Paris. These looted works, including the Rosetta Stone, Apollo Belvedere (Fig. 4.7), and the Horses of San Marco, are housed in individual niches. Each niche contains historical prints or photographs, as well as a pedestal with identifying information and a red button. Pressing the button for each work triggers a brief audio clip that provides additional context and insights into the artwork being viewed. The visual representations of these works are based on JPEG images from Wikimedia Commons, all of which are in the public domain. The examination of colonization of Africa follows, highlighting the case of the Benin Bronzes (prior to a number of nations in 2023 returning them). The 1907 Hague Convention is presented to offer context for understanding the consequences of such activities. While the convention addressed the prohibition of plundering during armed conflicts, it failed to explicitly address the protection of cultural heritage or artworks.

After successfully collecting all the works from the north side of the hall and placing them in the repository, the player is directed to the south side to explore the events of the World War II. Here, a concise overview of the Holocaust and the looting of an astounding 650,000 artworks by the Nazis across Europe is provided. Historical photographs lining the back wall offer context on the planned Führermuseum in Linz, Austria, which was intended to house an extensive collection of artworks, deemed the greatest on the continent. The player encounters examples of looted works, some stored in salt mines, and others tragically lost, such as Vincent van Gogh's *Painter on His Way to Work* (1888). Although van Gogh's work was never recovered, the player learns about the valiant efforts of the Monuments Men (1943–51), a group of American and British curators, architects, artists, art historians, and librarians who tirelessly traveled across Europe, attempting to catalogue and recover looted or stolen works by the Nazis (Saxon and Buchanan 2021). The legal battles to reclaim works

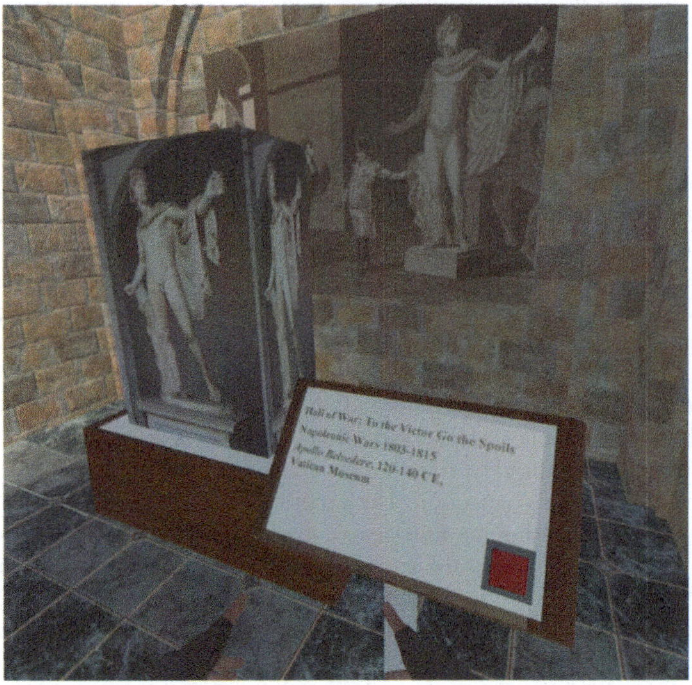

Fig. 4.7 Level 1: Hall of War: Apollo Belvedere. The Museum of the Lost, 2022. Lindenwood University. By permission of author

taken from Jews across Europe are then discussed, highlighting the cases of Klimt's *Portrait of Adele Bloch-Bauer* (1907), seen in the niche in Fig. 4.8, and Schiele's *Portrait of Wally Neuzil* (1912). Following the narrative in the north hall, the 1954 Hague Convention is also discussed, providing further insight into cultural heritage law. This convention further strengthened the 1907 Hague Convention by explicitly addressing protections for cultural heritage during armed conflict.

With the completion of the collection in "The Hall of War," a UI element triggers a quiz that the player must complete before progressing to the next level. The quiz is thoughtfully designed to be randomized, offering multiple questions for each topic covered to ensure a unique learning experience for different players. By integrating historical facts, ethical dilemmas, and game mechanics, this level fosters engagement and empathy while encouraging critical thinking about the complexities of cultural heritage protection and repatriation during wartime.

Level 2, titled "The Hall of Ownership," delves further into the intricate realm of cultural heritage law. Situated along the north wing on the second level of the museum, the level begins with the player ascending a staircase from Level 1. Upon reaching the top, the player is greeted by an enlarged still from the movie *Black Panther* (2018), depicting the character Killmonger observing African artifacts in a fictitious British museum. As the player spawns in, an audio clip automatically triggers, presenting the arguments advocating for repatriation, with the looting of Black culture from Africa

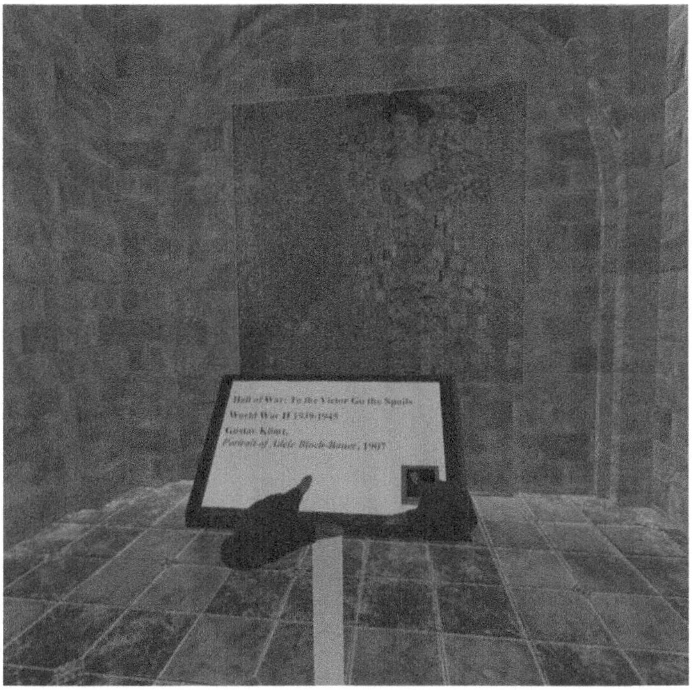

Fig. 4.8 Level 1: Hall of War: Portrait of Adele Bloch-Bauer. The Museum of the Lost, 2022. Lindenwood University. By permission of author

serving as a poignant backdrop. These arguments include the moral obligation to return cultural patrimony to rightful owners, emphasizing that cultural objects belong with the cultures that created them, as they are an integral part of their political and cultural identity. The level also highlights how the failure to return such objects can perpetuate colonialist ideologies, and how the geographical concentration of global museums in the industrialized North renders them inaccessible to the cultures that created the exhibited works.

Navigating along the corridor, the player encounters stacked crates serving as obstacles and the guards, who are now physically present and visible seen in Fig. 4.9. To progress, the player must deftly avoid the guards using stealth mechanics (ducking behind crates or peering around corners), making the experience more immersive and challenging. As the player progresses further, the case of the Elgin Marbles (Phidias (?), Parthenon sculptures, frieze: 438–432 B.C.E., pediment: c. 438–432 B.C.E., and metopes: c. 447–32 B.C.E.) is presented as an illustrative example where Greece has demanded the return of the works, and the aforementioned arguments for repatriation are considered. Through this level, players are prompted to critically reflect on the complexities of cultural ownership, the ethical dimensions of repatriation, and the significance of cultural identity in the context of museum collections.

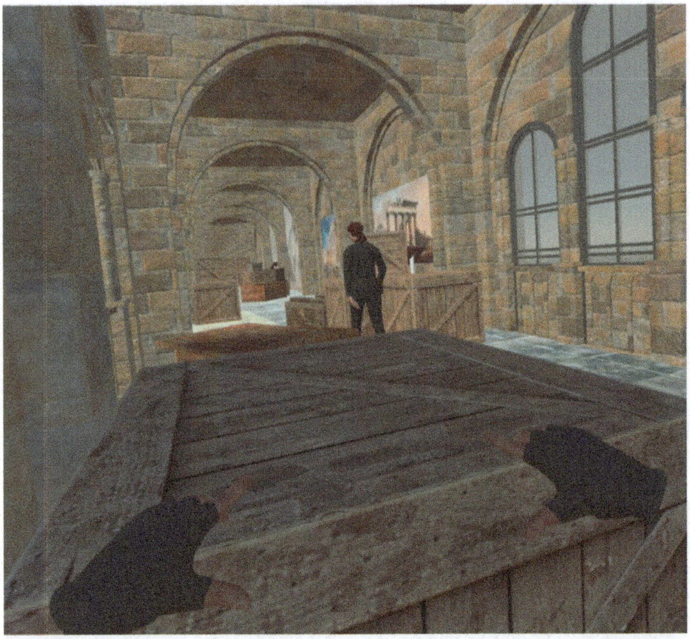

Fig. 4.9 Level 2: Hall of Ownership. The Museum of the Lost, 2022. Lindenwood University. By permission of author

The halfway mark of the north hall showcases the "Declaration on the Importance and Value of Universal Museums" of 2002, the response by 18 global museums across the world to the demand of repatriation of works in their collections. The arguments against repatriation are thus presented, include the argument that over time, the works in question have become part of the museum and nation that they are presented within and that often the peoples, nations, and political entities that created them no longer exist. Two examples follow—one in which repatriation was not carried out and one in which it was. The first is the case for the return of the *Bust of Nefertiti* (1341 B.C.E.) to Egypt. Smuggled out of the country in 1913 by German archaeologists and displayed in Berlin in 1924, Hitler refused Egypt's demands for *Nefertiti*'s return in 1933, stating instead that he would build an Egyptian museum to house the sculpture. The second, the *Sarpedon Krater* purchased by the Metropolitan Museum of Art in New York, was found to have been illegally looted from an Etruscan tomb in 1971. The laws that forced the return of the *krater* are then discussed at the end of the hall. The 1970 UNESCO Convention, for instance, allows for stolen or looted objects to be legally seized if there exists proof of ownership and the 1995 UNIDROIT Convention on Stolen or Illegally Exported Cultural Objects, demands the return of illegally excavated and/or exported cultural artifacts.

Upon progressing through Level 2, the player enters the south hall, which is thematically dedicated to instances of theft directly from art museums or works taken in situ, aptly titled "The Hall of Looting: Smash and Grab." This section offers a stark contrast to the previous levels, as it focuses on artworks stolen through audacious art heists carried out by individuals. Notably, these incidents do not involve wartime looting or contentious excavations but rather bold thefts directly from museums. The player first encounters the infamous art heist that took place on the night of March 18, 1990, at the Isabella Gardner Museum in Boston, Massachusetts. In what is considered the largest art heist in history, thieves made off with 13 valuable works, including Vermeer's *The Concert* (1664) valued at $250 million, and Rembrandt's *The Storm on the Sea of Galilee* (1633) (unknown value). Next, the Munch heists are presented, starting with the daring daylight theft of Munch's iconic painting *The Scream* (1893) valued at $110 million from the National Gallery in Oslo on May 7, 1994. Interestingly, the same work, along with Munch's *Madonna*, was stolen once again in 2004 (both works were recovered two years later). Finally, the player comes across Caravaggio's *Saint Jerome Writing* (1607) valued at $30 million, which was brazenly cut out of its frame in St. John's Co-Cathedral in Valetta, Malta. The artwork's absence serves as a stark reminder of the audacious nature of art theft.

Having collected all the works from both "The Hall of Ownership" and "The Hall of Looting," the player proceeds to the next stage, and similar to previous levels, a quiz is triggered, prompting players to demonstrate their knowledge of the topics and laws covered throughout Level 2. As before, the quiz questions are randomized, ensuring a varied experience for different players. Through this level, players are exposed to the complex issues surrounding art theft and the legal challenges associated with recovering stolen artworks.

The final level of the game, titled "The Hall of Stolen Fame," pays tribute to one of the most renowned artists in the Western art canon, Leonardo da Vinci, while shedding light on the impact of art theft on the fame of certain works. In this level, the player spawns at the base of the stairs on the lower level, west side, and is faced with seven works attributed to Leonardo. However, only one of these works has actually been stolen. The level is designed as a maze-like corridor of stacked crates that create a challenging environment for the player to navigate seen in Fig. 4.10.

Throughout the maze, the player encounters various works representing different stages of the artist's career, including *The Madonna of the Rocks* (1483–86), *Lady with an Ermine* (ca. 1489–91), *Madonna and Child with St. Anne* (ca. 1501), *Baptism of Christ* (ca. 1475), *Adoration of the Magi* (1481), and *La Belle Ferronniere* (1490–1499). Finally, at the end of the maze, the most famous work by Leonardo, the *Mona Lisa* (ca. 1503), awaits the player. As the objective is to identify the stolen work among the collection, this level poses a unique challenge. The educational strategy here relies on the recognition on the part of the player of the *Mona Lisa* as the most renowned piece in Leonardo's oeuvre, while the other works may be largely unknown to the general public. Each niche containing a work offers an audio clip and information about the piece. However, unlike the previous levels, the *Mona Lisa* is not housed in a niche but is featured prominently at the end of the maze. Historical photographs on display provide insight into one of the most famous art heists in history. The *Mona Lisa*, one of the most valuable paintings in the world, was stolen by Vincenzo Peruggia in 1911 and recovered in 1914. Ironically, Peruggia believed

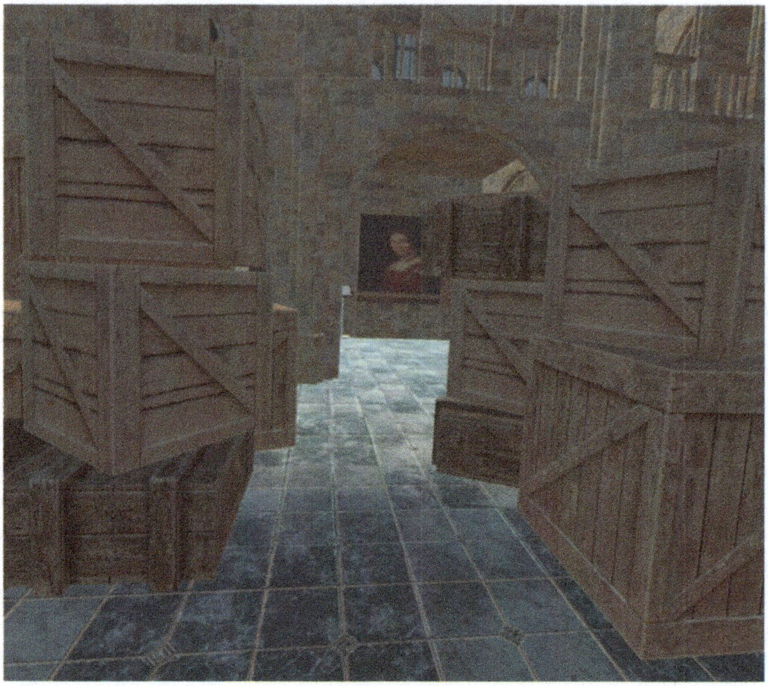

Fig. 4.10 Level 3: Hall of Stolen Fame. The Museum of the Lost, 2022. Lindenwood University. By permission of author

the work to have been looted by Napoleon during the Napoleonic Wars instead of being gifted by Leonardo to King Francis I (Oliveri et al. 2022).

Once the player deposits the final work in the repository, a final quiz awaits. However, unlike the earlier text-based quizzes, this one takes the form of videos with three distinct scenarios for the player to act upon. Each scenario involves an ethical dilemma concerning the actions the player chooses to take with the collected works. The player has the agency to decide whether to (1) sell the works in the illegal sales market, (2) return the works to the nation or people that created them, or (3) return them to the museum from which they were stolen. Each choice triggers a video showcasing potential outcomes, including being arrested, witnessing the joyous return of the works to their country of origin, or experiencing the general public's enjoyment of a universal museum experience. The narrative and user interface elements of the game deliberately leave the nature of the player's actions open-ended, allowing for magnanimous, selfish, or pragmatic choices. This final level of the game provides an introspective and thought-provoking conclusion, compelling players to consider the ethical complexities surrounding cultural heritage and the consequences of their decisions.

By gamifying the learning experience and leveraging the immersive capabilities of VR, the game not only engages players in active participation but also encourages critical thinking about the pressing issues facing the preservation of international cultural heritage. The three main levels of the game, "The Hall of War," "The Hall of Ownership," and "The Hall of Looting," each shed light on different aspects of art theft and its historical contexts, highlighting the urgent need to address these issues. Moreover, the final level, "The Hall of Stolen Fame," presents players with moral choices, offering agency and emphasizing the inherently political nature of art and its ownership. Art theft and the repatriation of looted works are complex matters intertwined with cultural identity, colonialism, geopolitical dynamics, and international law. The game's innovative approach exemplifies how immersive technology and game-based learning strategies can be harnessed to create meaningful educational experiences that challenge players to grapple with the multifaceted issues surrounding art and its global significance.

4.6 Beyond Postmodern Automatons: Politicizing Preservation

As highlighted in the gameplay of *The Museum of the Lost VR*, cultural heritage preservation is far from a neutral or apolitical endeavor; rather, it is deeply intertwined with politics and power dynamics. The concept of "Postmodern Automatons" by Rey Chow (1992) sheds light on this phenomenon, highlighting the mechanistic and dehumanized aspects of contemporary society that influence decisions regarding cultural heritage (Butler and Scott 2013). In this context, preservation efforts must be seen as inherently political, shaped by complex interactions between various actors and interests. The notion of "automatons" signifies how individuals and institutions are influenced and directed by external forces, social structures, and ideologies, leading to predetermined attitudes and behaviors. When it comes to the preservation of cultural heritage, considerations of value, significance, and ownership are not solely determined by objective measures but are also heavily influenced by political, economic, and cultural factors. Acknowledging this politicization is essential for understanding the complexities and challenges inherent in safeguarding cultural heritage in a globalized and ever-changing world. Beyond being a mere exercise in conservation, cultural heritage preservation becomes a critical space for examining the wider socio-political dimensions that shape our understanding of history, identity, and power relations. Therefore, this section will explore the implications of politicizing preservation efforts and how embracing this perspective can lead to more informed and conscientious approaches to safeguarding and repatriating cultural artifacts.

The analysis of the political ramifications of postmodern text by Chow transcends the simplistic framework of "modernism-postmodernism" as a linear progression or a binary opposition. Instead, Chow presents a more intricate perspective, where these concepts form a complex network of ideas, concepts, and artistic legacies, each evolving at different stages. Furthermore, Chow emphasizes that the implications of modernism and postmodernism are intricately tied to real historical and cultural contexts, taking into account factors such as time, location, gender, class, and social development. While her analysis primarily delves into the inherently political nature of feminist discourse, its relevance extends to the realm of cultural heritage preservation as well. Just like the treatment of feminist ideas, decisions regarding cultural heritage preservation are not solely shaped by abstract principles but are deeply rooted in historical, cultural, and political contexts. The notion that what constitutes modern or postmodern is contingent upon various socio-cultural factors underscores the need to view preservation efforts through a political lens, considering the complexities and power dynamics inherent in the safeguarding and repatriation of cultural artifacts.

The discussion revolves around the multifaceted understanding of modernism, its relationship to globalized terms like "modernity" and "modernization," and the technologization of culture. The focus on the visual realm, including the development of technologies like photography and film, expands the concept of the "visual" beyond mere physical vision. The dominance of the visual discourse in modernity reveals inherent epistemological issues in social relations, shaping constructions of social differences based on class, gender, and race. The traditional narrow view of modernism as a European and North American cultural awakening needs to be contextualized within a broader understanding of modernity's cultural expansionism with Eurocentric and patriarchal foundations. The postmodern era challenges and displaces modernism, and there is a confusion between modernism as a theoretical determinant and as a social effect. The postmodern critique disparages modernism, considering it mythical or narrative-driven, leading to a rewriting of history. Postmodernism challenges the clear cognitive boundaries set by modernism, with talk of dissolving boundaries and interchangeable positions of the observer and the observed. The proliferation of discourse in the postmodern era and the acceptance of diverse perspectives associate postmodernism with a sense of abandonment of previous limitations.

The discussion of the modernism-postmodernism problematic also raises broader questions about the politics of culture in general, including its intersection with feminism. The "displacement" of modernism by postmodernism is not simply a linear progression but a complex and multilayered phenomenon. For cultures beyond the dominant Western axis, modernism may be perceived as a foreign body, an omnipresent force whose physicality is an unavoidable reality. On the other hand, in the Western context, modernism still holds sway as an ideological legacy, a habit, and a familiar way of seeing. In this postmodern era, where boundaries appear to dissolve and anything seems permissible, postmodernism is sometimes seen as a belated articulation of what the West's "others" have been living all along. However, the relationship between feminism and postmodernism has not been without challenges. While feminists share the postmodern project of dismantling claims of cultural authority, their commitment to political struggles against the subordination of women complicates their acceptance of postmodern "universal abandon." For some feminists, the destabilization of conceptual boundaries and concrete beliefs may be perceived as a threat to their agenda of social progress based on reason and self-assertion.

The relationship between feminism and the postmodern scene shows a tension between embracing postmodernism's deconstructionist approach and preserving feminism's local and critical regional focus. This tension reflects the broader politics of culture, as it navigates the changing landscapes of modernism, postmodernism, and feminism. Understanding culture's displacement and transformation requires considering its interaction with power, identity, social development, and specific historical contexts. By recognizing these complex forces, scholars and activists can better understand and navigate the intricate world of cultural politics, postmodern identities, and feminist struggles in today's society. This understanding highlights the need to politicize the arts and cultural heritage, as decisions regarding preservation and representation are deeply intertwined with broader political, economic, and ideological factors. Taking a political approach to cultural heritage allows us to critically examine its complexities, power dynamics, and the impacts of historical and contemporary actions on various cultures and communities. As such, embracing the intersection of feminism, postmodernism, and culture can illuminate the importance of politicizing the arts to ensure inclusive and equitable representation and preservation.

The convergence of feminism and the postmodern milieu underscores a palpable friction: On one hand, there's an inclination toward the deconstructionist spirit of postmodernism; on the other, there's a pressing need to safeguard feminism's inherent regional critiques and localized political agendas. This tension is further complicated by the challenges posed to establishing a universally accepted global feminism, as differing cultural, social, and political landscapes can lead to disparate feminist narratives and priorities. The quest for a unified global feminist movement must navigate these intricate terrains, acknowledging diverse experiences while striving for common ground. This intricate relationship highlights the broader politics of culture in navigating the shifting terrains of modernism, postmodernism, and feminism. To understand the need for both political and technological strategies for digital cultural heritage preservation, lessons can be drawn from scenarios of destruction and reconstruction caused by the Taliban and ISIS, as well as the Russian-Ukraine War of 2022.

From 1996 until 2001, the extremist ideology of the Taliban, a predominantly Pashtun, Islamic fundamentalist group, led to the destruction of several giant Buddha statues on the ancient Silk Road in Afghanistan. In March 2001, the Taliban used explosives to demolish the statues, seen in Fig. 4.11, which had stood for over 1500 years, highlighting the critical need for preservation efforts to safeguard against such losses. Now, in a paradoxical twist, the Taliban charges tourists to show them the rubble left from their own demolition, a disheartening reminder of the consequences of neglecting cultural heritage preservation. Likewise, ISIS, the terrorist militant group that seized control of large areas of the Middle East became notorious for their destruction of cultural artifacts in Iraq and Syria beginning in 2014, causing irreparable damage to numerous historical sites, including ancient temples and archaeological sites. Their actions highlighted the urgency of digital preservation to document and reconstruct these invaluable cultural treasures. In response, digital reconstructions of the destroyed sites have been undertaken to record these

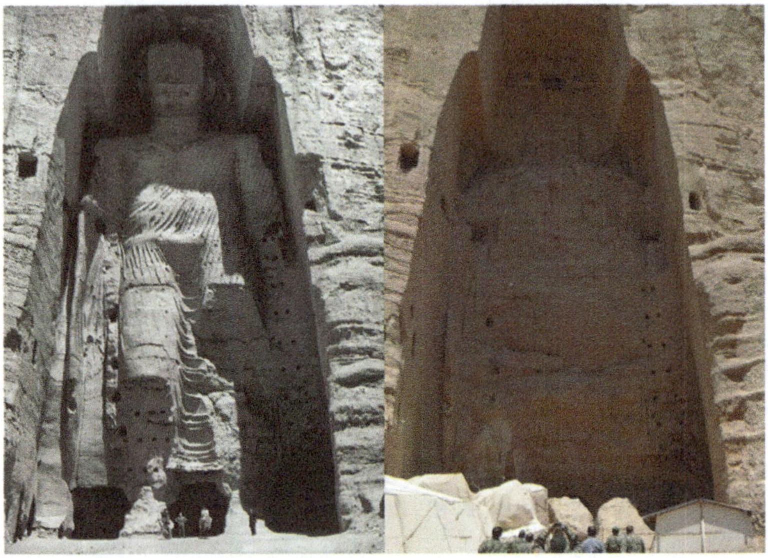

Fig. 4.11 Buddha of Bamiyan after destruction by Taliban in 2001 (from Wikimedia Commons, licensed under CC0)

historical places and underscore the importance of protecting cultural heritage from such devastation.

The Russian-Ukraine War of 2022 has instigated significant destruction, notably impacting Ukrainian cultural heritage sites seen in examples like Fig. 4.12. Such devastations in wartime bring to light stark differences in motivations and outcomes. In the case of ISIS, their targeting of cultural heritage had twofold motivations: firstly, to obliterate religious structures and symbols they deemed offensive and secondly, to loot and monetize valuable antiquities, channeling funds into their extremist operations. Contrastingly, Russia's aggression toward Ukrainian cultural heritage stems from a different ideological stance. Putin's refusal to acknowledge the distinct cultural identity of Ukraine, viewing it merely as an extension of Russia, translates into attacks that aim to erase Ukrainian heritage. This negation of identity is evident in the deliberate targeting of museums, libraries, monuments, and religious sites across Ukraine. The varying motivations and methods behind these attacks underscore the complexities and urgency of preserving global cultural heritage. To address these challenges, a combination of political and technological strategies is essential. Local communities and stakeholders have been involved in some efforts to garner popular support and instill a sense of ownership over their cultural heritage. Additionally, certain training and educational programs have been initiated to equip communities for active participation in preservation. However, these endeavors, while commendable, are merely initial steps. A more expansive and concerted approach is imperative to ensure the comprehensive protection and appreciation of cultural heritage.

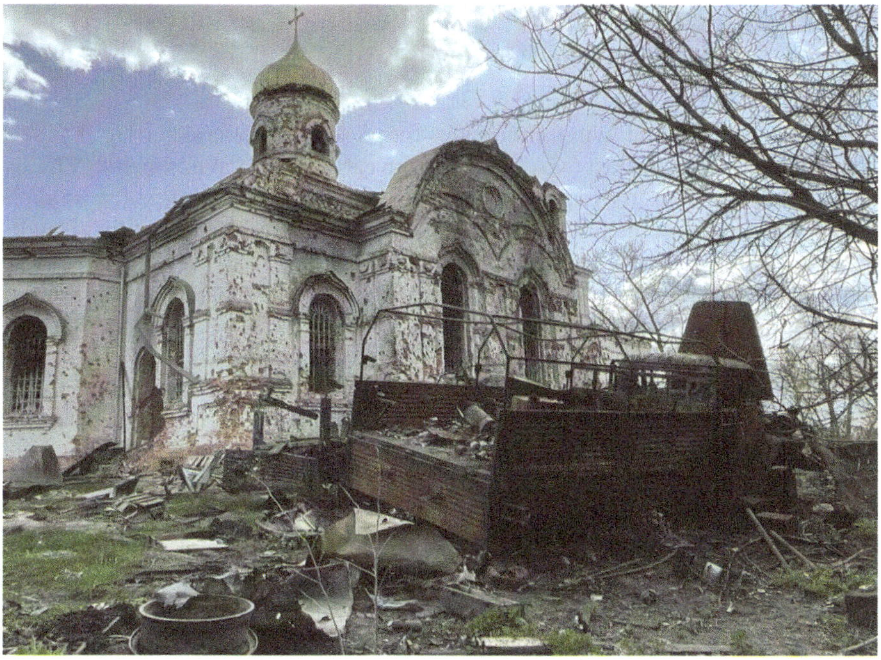

Fig. 4.12 Church of the Ascension (Lukashivka) after Russian attack, May 4, 2022 (from Wikimedia Commons, licensed under CC-BY 4.0)

On the technological front, developing user-friendly interfaces and accessible technologies is vital. This will enable a broader range of individuals, including non-experts, to participate in the digital preservation of cultural heritage. The ease of use will encourage more people to contribute to documenting and safeguarding cultural artifacts. Interoperability of equipment and file formats is another critical consideration. By standardizing formats and ensuring equipment compatibility, different organizations and institutions can collaborate more effectively, creating a more comprehensive and cohesive preservation network. This interoperability will allow for the seamless exchange of data and resources, enhancing preservation efforts.

The scenarios of destruction and reconstruction caused by the Taliban, ISIS, and the Russian-Ukraine War underscore the pressing need for a comprehensive approach to cultural heritage preservation. By combining political engagement, technological advancements, popular support, and training, the protection and accessibility of our global cultural heritage for future generations can be protected. Emphasizing the importance of both political and technological strategies will strengthen our collective efforts to preserve cultural heritage, safeguarding it from the devastating impacts of conflict and neglect.

4.7 Enabling Cultural Stewards: Building Capacity for Global Heritage Preservation

The utilization of digital twin technologies for preserving cultural heritage has seen significant growth in the last two decades. From early virtual environments (VE) and digital reconstructions that required complex workflows and multiple software applications and hardware, the field has advanced to now incorporate 3D artificial intelligence (AI) generative content and the latest photogrammetric scanning techniques. These advancements have made it easier for non-specialists to create digital twins of cultural artifacts and sites. Nevertheless, the urgency to preserve cultural heritage has intensified due to geopolitical instability, as seen in examples like the Russian invasion of Ukraine in 2022. Despite the availability of user-friendly and commercially accessible technologies, the digital art history and digital humanities fields are in a race against time to train and empower enough individuals on-site to create digital twins before more invaluable cultural artifacts and sites are lost to natural disasters, exacerbated by climate change, or destroyed in armed conflicts. However, the lack of international standards for methodological reproducibility remains a challenge, and many scholars still rely on specialized training and knowledge for their preservation efforts. In this section, we will explore how digital cultural heritage preservation can be scaled and easily taught using existing technologies and present a case study to demonstrate its practical implementation.

The adoption of immersive reality for cultural heritage preservation and reconstruction has rapidly increased, extending its reach beyond specialists and technicians to the general public, scholars, and non-specialists. Notably, the use of 3D scanning and photogrammetry has allowed for the digitization of lost or damaged interiors and the creation of 3D models of objects. For instance, the sculptural collection in the Uffizi Galleries in Florence, Italy, was captured individually through photogrammetric scanning and imported into the Unity game engine to craft an immersive experience (Brennan and Christiansen 2018). These technologies are now on the brink of combining their capabilities to reconstruct historically accurate interior and exterior architectural spaces, as well as archaeological sites, featuring the original furnishings and accoutrements that may have been dispersed or lost over time. The convergence of these technologies offers immense potential for creating compelling and authentic digital twins of cultural artifacts and sites, allowing for broader accessibility and engagement with cultural heritage.

The significance of three-dimensional virtual models in digital twins for cultural heritage preservation cannot be overstated. Their wide application in various industries, such as optimizing repair and maintenance schedules for physical plants, demonstrates their utility. Similarly, these digital assets have immense potential in monitoring the condition of heritage sites and artifacts in real-time. Digital twins enable the tracking of crucial environmental factors, such as temperature, humidity, and other variables that may impact the stability of cultural objects (Ariesen-Verschuur et al. 2022). By leveraging digital twins to predict the conditions, conservators and curators can proactively respond to environmental changes, ensuring the long-term preservation of our cultural heritage. These digital tools not only provide valuable insights into the present state of heritage sites and artifacts but also pave the way for innovative conservation strategies that can protect them for future generations.

The potential benefits of using digital twins in cultural heritage preservation and conservation should drive widespread adoption. By employing digital twins, conservators can proactively identify risks and implement preventative measures to safeguard objects and sites. Moreover, digital twins enable efficient planning and execution of restoration and conservation work, offering the flexibility to simulate various scenarios and select the most suitable approach for each specific artifact or site. Additionally, these reconstructions grant virtual access to heritage sites and artifacts, overcoming barriers that might limit physical visitation. This accessibility proves valuable for educational and curatorial purposes and can also benefit researchers by providing entry to understudied, historically significant locations that may be difficult or restricted for on-site exploration due to environmental, geopolitical, or security-related concerns. Furthermore, digital twins offer a solution for museums and collections seeking to enhance security measures in the wake of incidents such as the Just Stop Oil protests, which resulted in the defacement and damage of works in art museums. Overall, the use of digital twins in cultural heritage preservation has the potential to revolutionize conservation practices and broaden access to our shared human heritage. On the other hand, while the adoption of digital twins in cultural heritage preservation presents numerous advantages, certain drawbacks must be acknowledged. Primarily, an over-reliance on digital representations might inadvertently diminish the value and urgency of preserving the physical originals. Additionally, the creation and maintenance of these digital models demand substantial financial and technical resources, potentially sidelining institutions with limited budgets. Lastly, there are concerns about data security and the potential misuse of digital reconstructions in misleading or inauthentic presentations.

Advancements in technology have revolutionized the creation of digital twins for cultural heritage, streamlining the development process. In the past, 3D scanners and photogrammetry required laborious stitching of thousands of individual images by technicians. However, readily available systems like Matterport now generate interior renderings almost instantly, particularly for flat surfaces. The latest hand-held scanners, such as Scantech, can then produce high-resolution digital twins of objects with precise colors and textures. These systems are now portable, user-friendly, and affordable (ranging from $224 to $10,000). Furthermore, cutting-edge hardware and software enable polygonal rendering and low latency in the latest generation of headsets and game engines like Unity and Unreal Engine. This integration allows for a more authentic experience of cultural heritage by seamlessly combining architectural elements, spaces, and objects. The ongoing development of AI-powered 3D model generation further democratizes virtual content creation for cultural heritage, incorporating text prompts and 3D generative asset creation. These technological advancements hold immense promise in preserving and presenting cultural heritage in captivating and accessible ways, fostering a deeper appreciation for our shared past.

Building upon recent technological advancements, a scalable model is introduced, utilizing easily accessible and user-friendly 360 cameras and photogrammetry with LiDAR to capture cultural heritage sites. This model aims to quickly and effectively train non-specialists to create site-specific digital twins of diverse cultural heritage structures. An illustrative example of this user-friendly application is demonstrated through the detailed workflow of creating a digital twin of the Chiesa dei SS Apostoli e Biagio in Florence, Italy (Fig. 4.13) (accessible here: https://my.matterport.com/show/?m=sNCcHkVWmSb). The case study showcases how this approach, with minimal training, enhances access to cultural heritage sites, not only for research purposes but also for increasing public awareness of historical sites and museum collections and their interconnectedness. The proposed workflow model (Table 4.1), following the scalable framework by Hutson et al. (2023), demonstrates how institutions can employ digital twins to raise awareness of lesser-known historical structures that hold local significance and remain relatively unknown to the global scholarly community.

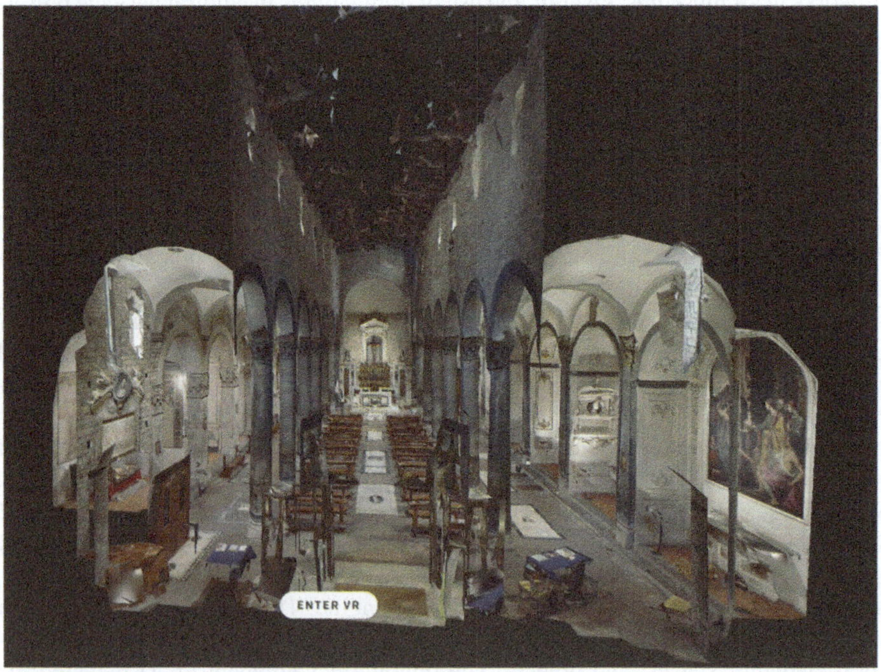

Fig. 4.13 Chiesa dei SS Apostoli e Biagio (overview of plan), Florence, Italy, begun 1075. Permission of author

Table 4.1 Best practices for cultural heritage digital twin creation

Phase 1	Steps
Preproduction	1. Perform reconnaissance on the location
	2. Familiarize with light sources and lighting conditions
	3. Ensure sufficient memory and charged equipment for travel
	4. Check weather conditions for optimal scanning
	5. Observe natural movements in the space for a realistic path
	6. Set up reference points and initial scan for orientation
	7. Plan scanning locations to create a smooth transition
	8. Take additional seated scans if needed
	9. Adjust and ensure scans are properly aligned and leveled
	10. Consider the contextual presence for embodiment
	11. Ensure the starting point is the entrance for natural flow
Phase 2	Steps
Production	1. Perform scanning process on location
	2. Adjust scans in real-time for accuracy and consistency
	3. Ensure all objects and surroundings are included in scans
	4. Take photos and capture the entire space
	5. Create a visual map of the museum adventure
	6. Provide options for interaction and information
	7. Allow curators to interact directly with visitors
	8. Incorporate elements for gamification and interactivity
	9. Take incremental scans for smoother transitions
	10. Take photos and create visual maps of the experience
Phase 3	Steps
Postproduction	1. Integrate content and metadata
	2. Provide a user-friendly homepage and layout view
	3. Design an intuitive and interactive user experience
	4. Include contextual information for understanding
	5. Embed additional content to enrich the experience
	6. Add hotspots and links for further exploration
	7. Create a seamless flow between scans and movements
	8. Ensure the curators are visible and interact with users
	9. Offer options for scaling and engaging audiences
	10. Design the museum experience based on user familiarity

The digital twin of the cultural heritage site can be accessed and experienced through various devices such as mobile phones, desktop computers, or VR in HMD. To initiate the creation of a digital twin, a series of steps must be undertaken, beginning with the preproduction phase. During this phase, the production team should conduct thorough reconnaissance on the location using existing maps, 360 photographs, or videos available through platforms like Google Maps or other geospatial records, if available. Familiarizing themselves with the location prior to the actual visit will enhance the efficiency of their work once on-site. During the reconnaissance, special attention should be paid to factors that may affect the quality of the digital twin. For instance, the team should assess the available light sources at the site, whether it is primarily lit by natural or artificial light, and whether additional lighting equipment needs to be considered for the scanning process.

During the preproduction phase of creating a digital twin for cultural heritage preservation, careful preparation and planning are essential to ensure a smooth and successful data collection process. The use of appropriate hardware and software plays a crucial role in capturing high-quality scans, and in the case study in Florence, the Matterport Pro 3D camera was utilized for the project. Before traveling to the location, the team must ensure that the hardware, including the camera, has enough memory to accommodate the size of the scanned files. It is essential to test the equipment prior to departure to ensure that everything is in working order. Adequate charging facilities, along with the necessary adapters and converters for the destination country's outlets and voltage, should also be prepared to avoid any disruptions during data collection.

Beyond the logistical preparations for data collection, building a collaborative rapport with local entities is imperative. Securing requisite permissions from local authorities to photograph or scan ensures adherence to legal and cultural norms. Additionally, transparently communicating the objectives, methodologies, and potential benefits of the documentation process to the custodians or operators of the church or cultural site fosters trust and collaborative spirit. Such proactive engagements not only streamline the data acquisition process but also respect and honor the sentiments and sovereignty of the communities hosting the cultural treasures.

Weather conditions are a crucial consideration during the scanning process. For the project using the pre-LiDAR Matterport system, overcast skies and diffused lighting were preferred to ensure optimal conditions for scanning and minimize disruptions caused by changing light sources. Once on-site, careful observation of how people naturally move through the space is necessary, particularly if the intention is to map out a path that reflects how a visitor would traverse the location in reality. This consideration is crucial in ensuring that the user experience in the digital twin closely aligns with the real-life experience. Mapping out worn paths and cutting corners can enhance the authenticity of the virtual experience.

Establishing a reference point, such as scanning the exterior of the structure, provides a natural orientation for the user and helps them understand the location within the larger context of the city or region. Paying attention to natural traversing of the interior, opening doors, and turning on artificial lights prior to scanning can further enhance the continuity of the digital twin experience. By carefully planning and executing the preproduction phase, the team can ensure that the data collection process for the digital twin is efficient, accurate, and provides an authentic representation of the cultural heritage site. With a clear understanding of the location and a well-thought-out workflow, the team can move on to the data processing phase to create an immersive and accessible digital twin for preservation, research, and public engagement.

During the production phase of creating a digital twin for cultural heritage preservation, meticulous attention to detail and adherence to a carefully planned workflow are crucial. Time should be factored on location, as the scanning process may take longer than initially projected. For example, in the case study, what was initially planned for two hours took four hours in total. When setting up the equipment, begin by setting the center of the aperture to the average height of the intended user to enhance the user's experience and embodiment. Ensure that the camera is level, as historical spaces may not be level or off-center. Resetting the center of the aperture height and re-leveling may be required for each scan to maintain accuracy.

To reduce VR motion sickness and create a well-crafted experience, pre-plan scanning locations to form a triangulation and grid of triangles. This triangulation will ensure smoother movements when navigating in the virtual space. When scanning, ensure that everything behind and around objects is included, allowing for a comprehensive view of the space and artifacts. For historical investigations, being able to see behind artifacts, such as altarpieces and in apses, can be valuable. During the scanning process, continuously check that everything is locating correctly with the hardware. Adjust scans in real-time if needed, as small variances can compound and disrupt processing in the cloud. Consistency in the scanning process is essential to avoid discontinuity of experience in the final digital twin.

An exception to the regimented scanning pattern is when there is a change in elevation, a doorway, or other funnel spots. In such cases, additional scans should be taken to more smoothly transition between locations, such as inside and outside of a structure. Moreover, it is essential to consider the purpose and unique features of the cultural heritage site during scanning. For example, in the case study, the team took seated scans to capture a live mass and embed the recorded file as a hotspot, adding an immersive and dynamic element to the digital twin. By carefully following these instructions and maintaining attention to detail during the production phase, the team can ensure the creation of a high-quality and authentic digital twin that accurately preserves the cultural heritage site and provides an engaging and educational experience for users.

After completing the scanning process, the postproduction phase begins, where the digital twin comes to life. Integration of content and metadata is crucial at this stage. The "fly-in" homepage should include the sponsoring organization's information and basic details in the dollhouse view (a zoomed-out view of the artifact/structure/site). User experience is of utmost importance in laying out the interior, aiming for the least number of clicks required to access desired information and experiences. The interior layout should be well-organized, with each area labeled for easy navigation. General information should be included to provide context, allowing users to understand their location when jumping between different areas. The front-facing and intuitive presentation should be the goal, making the overall experience seamless.

To ensure a smooth user experience, avoid overwhelming users with excessive information. Include essential contextual information in the first scene, and if needed, embed additional content to enable users to explore specific spots in more depth. Be cautious with external links, as they may take users out of the virtual experience. If linking to external sites, provide clear placeholders to inform users about the link's purpose and destination. Carefully consider the contextual presence for embodiment and ease of use. Select the entrance as the "home" scan reference point, mimicking how a real-world visitor would enter the location. Observations of how people move in reality should guide the virtual experience, making the entrance the natural starting point.

In following these postproduction guidelines, the team can ensure the digital twin provides an immersive and informative experience for users, preserving cultural heritage in an engaging and accessible manner. Properly executed, the digital twin will allow users to explore and appreciate historical sites and artifacts from anywhere in the world, fostering increased awareness, research, and appreciation of our shared cultural heritage. However, despite the progress in digital twin technologies for cultural heritage preservation, and examples provided here with the case study, challenges remain. Different countries and institutions are adopting various software, hardware, and development pipelines, hindering international collaboration due to issues of reproducibility, transferability, and interoperability. Moreover, the absence of an international organization providing interoperable standards for methodological reproducibility and file longevity poses a concern (Amato et al. 2022). Additionally, previously there remained specialized knowledge and training of technicians collaborating with scholars that impeded impede widespread adoption of their processes. At the same time, as these case studies demonstrate, there are a great deal of resources to assist art history and art and design faculty to integrate in coursework now.

Nonetheless, the use of digital twin technologies to preserve cultural heritage will undoubtedly advance with the adoption of streamlined processes outlined in this study. Current workflows for creating digital twins are laborious, expensive, and time-consuming. However, with the accessibility of 3D AI generative content and the latest photogrammetric scanning technology, even non-specialists can now create digital twins for historical research with minimal training. Previously restricted to high-end professionals, 3D cameras have become affordable and user-friendly, with the potential for quick training (as outlined earlier). This study exemplifies how technologies like Matterport, initially used in real estate and insurance industries, can be adapted to preserve cultural heritage. As such technologies become more widely adopted, international standards for methodological reproducibility may become feasible. Future advancements in 3D AI generative content will likely enable true six degrees of freedom (6DoF), overcoming current limitations, and offering enormous value to museums and cultural heritage institutions for quick and affordable creation of digital twins to safeguard irreplaceable historical sites and artifacts.

In all, the use of digital twin technologies in cultural heritage preservation has seen remarkable progress in recent years. From virtual environments and digital reconstructions, we have advanced to the immediacy of 3D artificial intelligence generative content and photogrammetric scanning, enabling even non-specialists to create digital twins for historical research. However, to fully harness the potential of these technologies, challenges such as interoperability, methodological standards, and specialized training need to be addressed on an international level. To move forward, it is crucial for the cultural heritage preservation community to come together and establish global standards for methodological reproducibility and file longevity. UNESCO had used a file sharing system for European projects, but an international standard is still needed. An international organization should be formed to ensure interoperability and facilitate collaboration among institutions and countries. Moreover, there should be a concerted effort to provide prepackaged training programs to upskill non-specialists worldwide, enabling them to assist in digital twin creation.

New developments are making this a reality: Esri's global digital twin capability, along with the ability to overlay external scans and incorporate BIM files, will revolutionize the process of digital twin creation. By making scanning accessible through commonly available devices like smartphones and using LiDAR technology to add point clouds to Esri, historical reconstruction can become possible from anywhere in the world, even without specialized training. Imagine if this technology had been available during events like the US pullout from Iraq or the onset of the Ukraine-Russia War—countless cultural artifacts and structures could have been preserved for future generations. The call to action is clear: Global cooperation is necessary and the adoption of standardized methodologies, along with training provided to non-specialists. With these efforts, the full potential of digital twin technologies in cultural heritage preservation can be unlocked, safeguarding the world's rich history and cultural identity for generations to come.

References

Amato A, Aversa R, Branco D, Venticinque S, Renda G, Mataluna S (2022) Porting of semantically annotated and geo-located images to an interoperability framework. In: Computational intelligence in security for information systems conference, June 2022. Springer International Publishing, Cham, pp 508–516

Ariesen-Verschuur N, Verdouw C, Tekinerdogan B (2022) Digital twins in greenhouse horticulture: a review. Comput Electron Agric 199:107183

Auerbach JA, Hoffenberg PH (eds) (2008) Britain, the empire, and the world at the great exhibition of 1851. Ashgate Publishing Ltd.

Barbosa MG, Saboya LAD, Bevilaqua DV (2021) A survey and evaluation of mobile apps in science centers and museums

Bekele MK, Pierdicca R, Frontoni E, Malinverni ES, Gain J (2018) A survey of augmented, virtual, and mixed reality for cultural heritage. J Comput Cult Herit (JOCCH) 11(2):1–36

Ben Ghida D (2020) Augmented reality and virtual reality: a 360 immersion into western history of architecture. Int J 8(9)

Biocca F, Levy MR (2013) Communication in the age of virtual reality. Routledge

Brendemoen B (2020) The eighth international congress of orientalists, held in Stockholm/Uppsala and Christiania (1–14 September 1889), and its echo in Turkish literature. In: Turcologica Upsaliensia. Brill, pp 129–144
Brennan M (2018) Virtual reality in the humanities classroom. University of Indiana, Bloomington
Brennan M, Christiansen L (2018) Virtual materiality: a virtual reality framework for the analysis and visualization of cultural heritage 3D models. In: Digital heritage 2018, 3rd edn
Buchhart D, Montez R, Ricard R, Yablonksy L, Warsh L (2022) Keith Haring | Jean-Michel Basquiat: crossing lines. Princeton University Press
Butler J, Scott JW (eds) (2013) Feminists theorize the political. Routledge
Carrier JG (ed) (1995) Occidentalism: images of the west. Clarendon Press
Casu A, Spano LD, Sorrentino F, Scateni R (2015) RiftArt: bringing masterpieces in the classroom through immersive virtual reality. In: STAG, Oct 2015, pp 77–84
Champion E, Foka A (2020) Art history, heritage games, and virtual reality. In: The Routledge companion to digital humanities and art history. Routledge, pp 238–253
Corbillé S, Fantin E (2020) Les animaux dans les expositions universelles au xixe siècle: monstration, ordonnancement et requalification du vivant. Paris et Londres, 1851–1889. Cult Mus Muséol Recher Cult 35:211–241
Craven A (2012) Esmeralda of Notre-Dame: the gypsy in medieval view from Hugo to Disney. In: The Disney Middle Ages: a fairy-tale and fantasy past. Palgrave Macmillan US, New York, pp 225–242
Dastyar S (2019) The investigation of the effectiveness of participatory learning education on students motivation and academic achievement. Int J Adv Res Publ 3(8):165–170
Deterding S, Hook J, Fiebrink R, Gillies M, Gow J, Akten M et al (2017) Mixed-initiative creative interfaces. In: Proceedings of the 2017 CHI conference extended abstracts on human factors in computing systems, May 2017, pp 628–635
Drabinski JE (1994) The status of the transcendental in Levinas' thought. Philos Today 38(2):149
Dubois A (2023) Literature for all: on designation and interpretation in Hugo's Notre-Dame de Paris. Fr Stud 77(2):183–197
Elkins J (ed) (2007) Is art history global?, vol 3. Taylor & Francis
Froschauer J, Arends M, Goldfarb D, Merkl D (2011) Towards an online multiplayer serious game providing a joyful experience in learning art history. In: 2011 third international conference on games and virtual worlds for serious applications, May 2011. IEEE, pp 160–163
Gee JP (2011) Reflections on empirical evidence on games and learning. In: Computer games and instruction, pp 223–232
Godlewska A (1995) Map, text and image. In: The mentality of enlightened conquerors: a new look at the description de l'Egypte. Transactions of the Institute of British Geographers, pp 5–28
Gorman JM (2013) Universalism and the new museology: impacts on the ethics of authority and ownership. In: New directions in museum ethics. Routledge, pp 76–89
Gullino P, Larcher F (2013) Integrity in UNESCO world heritage sites. A comparative study for rural landscapes. J Cult Herit 14(5):389–395
Hendry J (2020) The orient strikes back: a global view of cultural display. Routledge
Huaman EMR, Aceituno RGA, Sharhorodska O (2019) Application of virtual reality and gamification in the teaching of art history. In: Learning and collaboration technologies. Ubiquitous and virtual environments for learning and collaboration: 6th international conference, LCT 2019, held as part of the 21st HCI international conference, HCII 2019, proceedings, part II 21, Orlando, FL, 26–31 July 2019. Springer International Publishing, pp 220–229
Hutson J, Fulcher B (2023) A virtual reality educational game for the ethics of cultural heritage repatriation. Games Cult 18(6):759–782
Hutson J, Weber J, Russo A (2023) Digital twins and cultural heritage preservation: a case study of best practices and reproducibility in Chiesa dei SS Apostoli e Biagio. Art Des Rev 11
Hyun H, Park J, Ren T, Kim H (2018) The role of ambiances and aesthetics on millennials' museum visiting behavior. Arts and the Market 8(2):152–167

Ivakhiv A (2015) The age of the world motion picture: cosmic visions in the post-earthrise era. In: The changing world religion map: sacred places, identities, practices and politics, pp 129–144
Kemp M (2007) Leonardo da Vinci: the marvellous works of nature and man. Oxford University Press
Kim S, Song K, Lockee B, Burton J (2017) What is gamification in learning and education? In: Gamification in learning and education: enjoy learning like gaming. Springer International Publishing, Cham, pp 25–38
Korallo L (2010) Use of virtual reality environments to improve the learning of historical chronology. Doctoral dissertation, Middlesex University
Larson K (2020) Serious games and gamification in the corporate training environment: a literature review. TechTrends 64(2):319–328
López-Martínez A, Carrera Á, Iglesias CA (2020) Empowering museum experiences applying gamification techniques based on linked data and smart objects. Appl Sci 10(16):5419
Makarova EA, Makarova EL, Korsakova TV (2019) The role of globalization and integration in interdisciplinary research, culture and education development. J Hist Cult Art Res 8(1):111–127
Martin E (2011) 'Ceci tuera cela'? Literary canons and the challenge of visual imagery and popular culture. In: The canonical debate today. Brill, pp 227–245
Mitchell T (2019) Orientalism and the exhibitionary order. In: Grasping the world. Routledge, pp 442–460
Moorefield-Lang H (2015) Libraries and the rift: oculus rift and 4D devices in libraries. Knowl Quest 43(5):76–77
Nasirdinova Y (2022) Central Asia through the prism of French oriental studies. Galaxy Int Interdiscip Res J 10(11):1017–1025
Oliveri V, Porter G, Davies C, James P (2022) Art crime: the challenges of provenance, law and ethics. Mus Manag Curatorsh 37(2):179–195
Prensky M (2003) Digital game-based learning. Comput Entertain (CIE) 1(1):21–21
Preziosi D (2009) The art of art history: a critical anthology. OUP, Oxford
Said E (1978) Orientalism: western concepts of the orient. Pantheon, Nueva York
Saxon EJ, Buchanan SA (2021) Recollecting the archives of artist and monuments Man Lewis Stanton Williams. Journal of Archival Organization 18(1-2):37–50
Schneeweis A (2016) Desired, yet unwanted: Hugo's gypsy bodies in popular media. Pop Commun 14(2):99–110
Shernoff D, Hamari J, Rowe E (2014) Measuring flow in educational games and gamified learning environments. In: EdMedia + innovate learning, June 2014. Association for the Advancement of Computing in Education (AACE), pp 2276–2281
Spiqel A (2006) Hugo, Victor: Victor Hugo au sacre de Charles X (1825). Biography 29(1):222–223
Squire K (2006) From content to context: videogames as designed experience. Educ Res 35(8):19–29
Steger MB (2003) Globalization: a very short introduction
Szwydky LL (2010) Victor Hugo's Notre-Dame de Paris on the nineteenth-century London stage. Eur Romant Rev 21(4):469–487
Van Eck RN, Guy M, Young T, Winger AT, Brewster S (2015) Project NEO: a video game to promote STEM competency for preservice elementary teachers. Technol Knowl Learn 20:277–297
Wideman HH, Owston RD, Brown C, Kushniruk A, Ho F, Pitts KC (2007) Unpacking the potential of educational gaming: a new tool for gaming research. Simul Gaming 38(1):10–30
Wolf MJ, Perron B (eds) (2013) The video game theory reader. Routledge
Zhang X, Zhang L, Ullah A (2022) Digital twin technology and wisdom museum from the perspective of artificial intelligence. In: International conference on multi-modal information analytics, Apr 2022. Springer International Publishing, Cham, pp 674–682

Part III
Identity and Interactivity

Chapter 5
Embodiment in Avatars and Technoculture

Abstract This chapter investigates how avatars, as digital representations of self, facilitate new forms of cultural expression and identity exploration within digital spaces. It discusses the role of avatars in embodying cultural, racial, and gender identities, and how these digital personas influence users' experiences and interactions in virtual realms. The narrative highlights the transformative potential of avatars in transcending physical limitations and fostering inclusivity by allowing users to experiment with and redefine aspects of their identity. The chapter also examines the socio-cultural implications of avatars in cultural heritage contexts, where they serve as mediators of historical narratives and cultural learning. By focusing on both the opportunities and challenges presented by avatars in cultural engagement, the text explores how digital embodiment impacts the perception of self and others in increasingly interconnected virtual societies. Through detailed analysis of user engagement and the performative aspects of avatars, this chapter contributes to a broader understanding of the digital landscape's influence on contemporary identity and cultural dynamics.

5.1 Hegemonic Avatars: From Divine Manifestations to Virtual Representations

The convergence of technology and culture has given rise to a new digital landscape where virtual realities, avatars, and interactive experiences redefine how individuals perceive and engage with art and cultural heritage. The revolutionary potential of digital embodiment provides unparalleled chances for engaging with artworks and exploring diverse histories from the perspective of entirely different identities, encompassing abilities, nationalities, genders, ages, races, and ethnicities. Avatars, in particular, enable us not only to inhabit the roles of other individuals but also to assume the personas of fictional or real characters, various species, objects, elements, and countless other entities, expanding the boundaries of our virtual experiences. With the ability to move between portals and change avatars, the very nature of identity is called into question as performative self-expression and the evolving role of

J. Hutson, *Art and Culture in the Multiverse of Metaverses*, Springer Series on Cultural Computing, https://doi.org/10.1007/978-3-031-66320-8_5

avatars in this virtual reality. Drawing inspiration from Judith Butler's seminal work on performative acts and gender constitution, the historical significance of avatars and their potential evolution in this boundless realm, where participants can embody any form shall be investigated (Butler 2020). In this technocultural exploration, the profound interplay between identity, art, and virtuality in the metaverse, shaping the understanding of self and artistic experience in unprecedented ways will be the primary focus.

When considering the term "avatar," the 2009 science fiction epic directed by James Cameron is often referenced as the film that returned the word to the modern lexicon. The movie is set in the mid-twenty-second century on the fictional planet of Pandora, which is inhabited by the Na'vi, a humanoid species deeply connected to nature. The story revolves around the human Jake Sully, a paraplegic former Marine, who participates in a corporate mission to exploit Pandora's resources by remotely controlling an avatar. With a genetically engineered alien-human hybrid created for him, the previously immobile Jake is able to control a body that not only allows him to walk again, but performs super-human, acrobatic feats. With the box-office success of *Avatar* (and the 2022 sequel), the film continues to shape a popular culture understanding of the term "avatar," which now often refers to a virtual representation of oneself used in digital environments, games, and social media, further reflecting the exploration of identity and embodiment in a technologically driven world. Interestingly, the embodiment in flesh as portrayed in the film is closer to the original meaning of the term than our modern usage.

The term "avatar" holds a rich historical significance that goes beyond its modern technological connotations. Its origins can be traced back to the Sanskrit word *avatāra*, meaning "descent" or "to descend" (Rosen 2023). Initially referring to the earthly manifestation or incarnation of a deity from Hindu or Buddhist religious traditions, avatars were primarily associated with the incarnation of the Hindu god Vishnu, or other divine beings, and believed to be manifestations of spiritual entities descending to the earthly realm with a specific purpose. In Hindu tradition, avatars hold a central role, embodying divine principles and undertaking significant tasks to restore balance and righteousness. For example, the ten avatars of Vishnu, known as the Dashavatara, include the iconic avatars such as Rama and Krishna. Each of these avatars serves a unique purpose, fulfilling their divine mission on Earth (Monga and Das 2021). Buddhist traditions also recognize the concept of avatars, known as Bodhisattvas. These enlightened beings delay their entry into nirvana to remain in the mortal world and assist others on their spiritual paths. Avalokiteshvara, the Bodhisattva of Compassion, is one such avatar revered across various Buddhist cultures.

The concept of avatars has transcended various mythologies and folklore in different cultures, revealing its universal presence and enduring significance. Beyond Eastern religious traditions, avatars find representation in ancient Egyptian mythology, where gods were often depicted with animal heads and human bodies, embodying their divine attributes and symbolism (Nyord 2020). These anthropomorphic representations served as incarnations of the gods, connecting the divine realm to the physical world. For instance, the Egyptian god Horus, seen in Fig. 5.1, often portrayed with the head of a falcon and the body of a man, represented the qualities of the god of protection and kingship. In ancient Egyptian mythology, gods and goddesses were believed to possess multiple forms, capable of appearing in various ways, be they human or animal. For example, the goddess Bastet could appear as a lioness, symbolizing her fierce and protective nature, or as a woman, representing her nurturing and motherly aspects (Abdallah et al. 2022). This fluidity of avatars allowed gods to interact with mortals in ways that were relatable and comprehensible to humans.

Turning to Greco-Roman mythology, there are numerous examples of gods transforming into avatars, as in the *Metamorphoses*, written in the 8 C.E. by the Roman poet Ovid. Written as a secular narrative poem, assembling fragmentary stories and hymns from the Greeks and other sources, in order to chronicle the history of the world from its creation to the deification of Julius Caesar, focusing on the theme of transformation and metamorphosis in various mythological stories. In one instance,

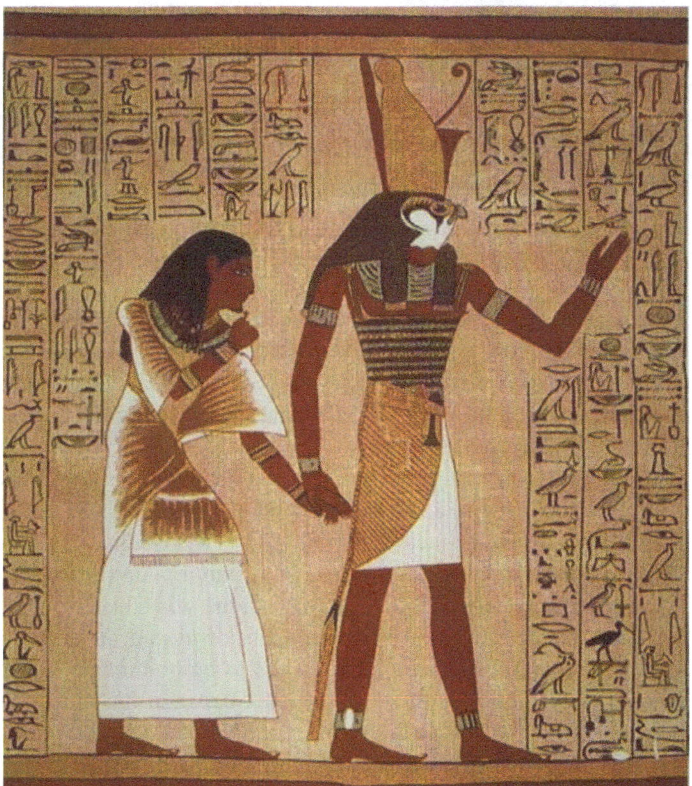

Fig. 5.1 Egyptian God Horus, thirteenth-century B.C.E. (from Wikimedia Commons, licensed under CC0)

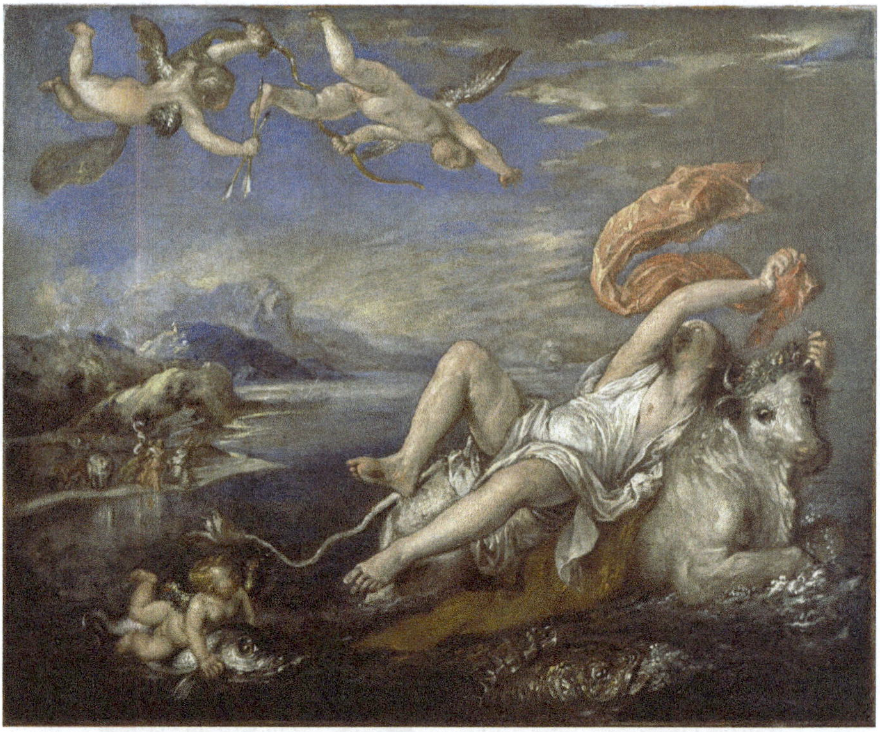

Fig. 5.2 Titian, Rape of Europa, ca. 1560–62 (from Wikimedia Commons, licensed under CC0)

the god Jupiter transformed into a bull to seduce the princess Europa on the island of Crete seen in Fig. 5.2, leading to the myth of the Minotaur, the half-man half-bull creature at the center of the Labyrinth of King Minos. Such transformations, or avatar manifestations, served as a means for gods to interact with humans and shape the course of their lives in unforeseen ways.

The influence of the concept of avatars extended to various Eastern Mystery cults at the end of the Roman Empire and eventually played a central role in Christianity. In Christianity, the belief in the incarnation of God as Jesus Christ, as illustrated in Fig. 5.3, echoes the traditional understanding of avatars in Hinduism. The idea that God took human form to dwell among mortals aligns with the essence of avatars in Hindu and Buddhist traditions. The universality of avatars in different cultural contexts demonstrates their enduring significance as embodiments of divine qualities, agents of transformation, and intermediaries between the mortal and divine realms. But these embodiments extend beyond the divine realm, playing a direct and central role in human cultures globally. In addition to the religious and mythological contexts, avatars find expression in various socio-cultural practices across different regions and time periods. African tribes, for instance, have employed the concept of avatars in their rituals to connect with the spiritual realm and affect the supernatural world. Some tribes adopt the persona of spirits, such as in the example of the

Tubuan Mask, which was worn by the Tolai people in Papua New Guinea during dance ceremonies. The wearers of the masks take on the identity of spirits to either scare away malevolent forces or bring about positive influences in their community (Tateyama 2016). Similarly, among the Native American tribes of the Pacific Northwest and Paleolithic cultures, shamans transformed into animals, exemplified by the Kwakwaka'wakw on Vancouver Island, Canada. The significance of transformation masks extends beyond their visual impact seen in Fig. 5.4. In the Kwakwaka'wakw mythology, moments of transformation involving trickster supernaturals, such as Raven and Thunderbird, play important roles. Raven, known as a consummate trickster, transforms into different creatures and provides useful items to humans, while Thunderbird causes thunder and lightning when he beats his wings. The masks themselves depict these transformative moments, allowing dancers to embody the spirit of the ancestors and gain insights into the spiritual world (Masco 1995). The transformation into animals was believed to enable shamans to connect with spiritual beings and gain insight into the spiritual world, facilitating healing and guidance for their communities.

The concept of avatars is not limited to specific regions or ancient cultures; it has also been present in Norse traditions. Viking berserkers, for example, adopted the persona of bears, wolves, or wild boars during battle. These warriors were legendary for their savagery and fierce frenzy in combat, fighting in a trance-like fury akin to the ferocity of animals. The term "berserker" is derived from the Old Norse word *berserkr*, meaning "bearskin," reflecting their association with bears. They believed that by transforming into these animals, they could harness their strength and courage in battle, exemplifying a form of avatar embodying animal attributes (Dale 2021). Furthermore, avatars have been represented in Native American cultures, particularly among the Pueblo tribes of the American Southwest. The Katsina dolls, created by the Hopi people, are believed to be physical embodiments of ancestral spirits, representing various deities and natural elements. These dolls are used in ceremonies and rituals to connect with the spirit world and invoke blessings, rain, fertility, and other aspects of life. In this context, the Katsina dolls serve as avatars, enabling individuals to establish a connection with their spiritual beliefs and ancestral heritage.

Throughout history, the concept of avatars has demonstrated that the multifaceted fascination humanity has with self-representation, transformation, and connection with the divine or supernatural, seeing themselves transcend time and space in their own visage. Whether through religious rituals, mythological tales, or artistic expressions, avatars continue to be a fundamental aspect of human culture, allowing individuals to explore diverse identities and engage with the spiritual and metaphysical realms. But the examples discussed above predate the Industrial Revolution, and secular societies, thus primarily discuss these embodiments in religious or spiritual terms. However, with the advent of the computer age and subsequent technological advancements, the concept of avatars took on a new dimension, becoming accessible and widespread due to their affordability, reproducibility, and adaptability. In the contemporary digital landscape, the term "avatar" has seamlessly transitioned

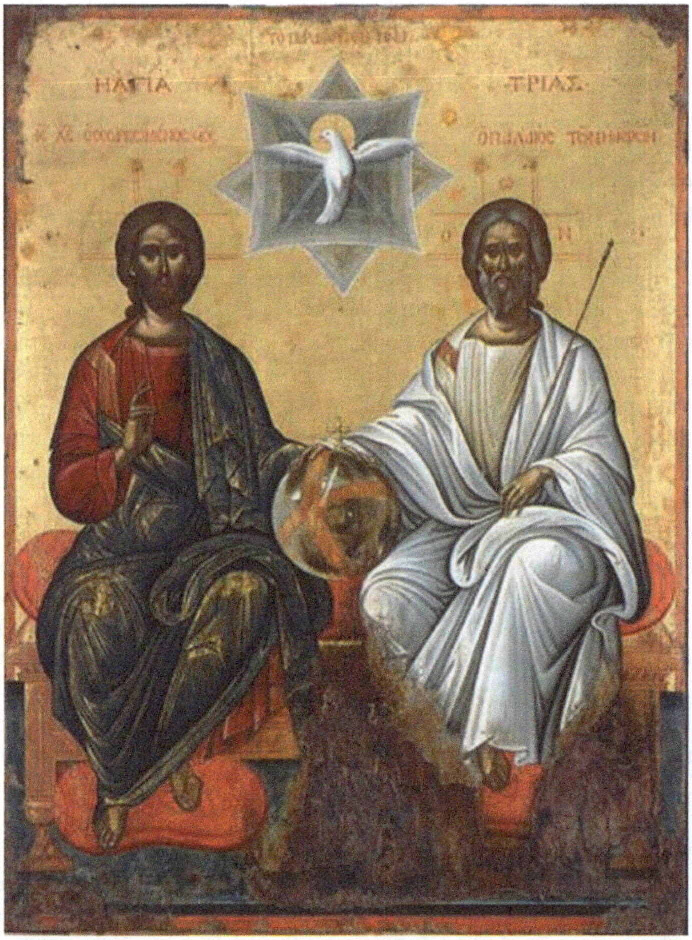

Fig. 5.3 Possibly Elias Moskos or Michael Damaskinos, The Holy Trinity, Cretan School, ca. 1550–1593 (from Wikimedia Commons, licensed under CC0)

into the realm of virtual spaces, where it continues to serve as a representation of a user identity or existence. While this technology-driven notion of avatars is relatively recent, its foundational idea of embodying diverse forms or identities traces back to profound historical and cultural origins that transcend civilizations and belief systems.

In the digital age, the notion of avatars has become a crucial part of *technoculture*—a term referring to the profound impact of technology on human societies—encompassing the intricate interplay between technology and culture, where advancements in technology shape the way people think, communicate, interact, and create significance in their lives. Technoculture influences various facets of modern existence, ranging from art and communication to education, industry, entertainment,

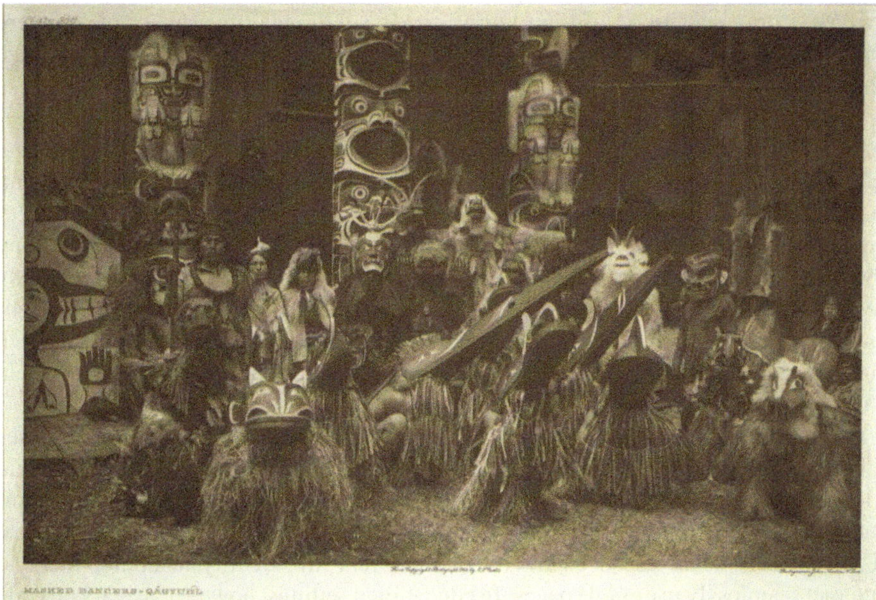

Fig. 5.4 Edward S. Curtis, Kwakwaka'wakw Potlatch, British Columbia, ca. 1914 (from Wikimedia Commons, licensed under CC0)

social norms, and individual self-identities (Octobre 2020). This dynamic relationship between technology and culture embraces both positive and negative consequences, while also fostering new cultural practices and attitudes in response to technological transformations. Understanding the concept of technoculture offers valuable insights into how technology becomes deeply integrated into society, molding human experiences, and influencing cultural values and norms.

In the realm of technology and digital media, an avatar takes the form of a graphical representation or virtual character, embodying a user in virtual environments like computer games, social media platforms, or virtual spaces. These avatars are highly customizable, allowing users to tailor them to their likeness, personality, or preferences. Acting as a digital alter ego or persona, avatars enable users to interact and engage with others within the virtual world (see Fig. 2.13). They come in various forms, ranging from simple 2D icons or profile pictures to more complex 3D models that can move and interact with the virtual surroundings. As online communities, social networks, and virtual reality applications continue to flourish, the concept of avatars gains significance, offering users a profound sense of presence and identity in the vast digital landscape.

In the contemporary landscape, avatars have found diverse applications in industry, training, and entertainment, offering unique opportunities for users to engage and interact with virtual environments. Notably, the use of interoperable avatars has revolutionized industries like telehealth with remote patient monitoring

(RPM). Companies like Vuforia Enterprise AR Solutions (https://www.ptc.com/en/products/vuforia) have developed platforms that allow users to create and customize their avatars, which can then represent them in virtual workspaces. With these avatars, remote teams can collaborate effectively, bridging geographical distances and fostering seamless communication (Benito and González 2019). These digital embodiments have also proved to be instrumental in training and development, particularly in honing "soft skills" like communication, empathy, and conflict resolution. Platforms like BodySwaps (https://bodyswaps.co/) utilize proprietary training systems to offer users realistic and immersive training scenarios. Users can embody avatars that replicate real-life professional situations, allowing them to practice essential interpersonal skills without real-world consequences. The ability to interact with these avatars in various situations helps users build confidence and proficiency in their communication and problem-solving abilities.

Avatars now come in a wide array of forms, each designed to suit specific purposes and contexts. In social and professional engagements, users often opt for avatars that closely resemble their own appearance, allowing them to project a digital representation of themselves in virtual meetings, conferences, or social media platforms. These avatars serve as an extension of their real-life identity, enabling them to interact seamlessly in digital spaces. In educational settings, however, avatars provide a realm of creative expression and learning opportunities. Students may have the freedom to choose avatars that are vastly different from their real-life personas, such as cartoon characters or animals (Nizam et al. 2022). This diversity fosters a playful and engaging atmosphere, making learning experiences more enjoyable and personalized.

Historically accurate avatars represent a fascinating fusion of education and entertainment. These avatars transport users to particular settings, geographical locations, and specific points in time, allowing them to experience historical events and cultural moments firsthand. Such applications are valuable for educational institutions, museums, and cultural heritage preservation efforts. Users can immerse themselves in the past, learning from an interactive and experiential perspective that textbooks cannot replicate. In all, avatars have become powerful tools that transcend industries and serve as vital components of modern-day technology. Their role in industry, training, and entertainment is continually evolving, unlocking new possibilities for human–computer interactions. Whether capturing personal likeness for professional engagements, offering creative expression in educational settings, or enabling historical exploration, avatars provide a transformative and enriching experience for users in the digital age.

5.2 Avatars and Performative Identity Negotiation in the Digital Age

Avatars play a significant role in negotiating individual and group identities, just as seen in historical examples (Kirby and Özkula 2023). This highlights the fluidity and adaptability of identity, which spans across time and is not fixed but rather enacted within socially acceptable boundaries. Identity encompasses the distinct qualities that define individuals or groups, shaping their self-perception and relationships with others and the world. Multiple factors contribute to identity formation, including cultural background, social roles, personal experiences, beliefs, values, and physical attributes. The process of identity development commences during the Concrete Operational Stage, where children begin to recognize the uniqueness of their thoughts and emotions, understanding that they may not be universally shared or even grounded in reality. In this context, the discussion will explore how avatars and performative identity negotiation further shape and influence the construction of identities in the digital age.

Pretend play and avatars offer crucial pathways for fostering identity development during early childhood in developmental psychology. These activities empower children to experiment with diverse roles, perspectives, and emotions, facilitating exploration of various aspects of self-identity and a deeper understanding of human interactions. Through imaginative engagement, children have the opportunity to don different personas and engage in self-reflection, thus contributing to the formation of their self-concept and self-awareness (Walker and Weidenbenner 2019). The active engagement with self-reflective activities assists in development of identity in general, which at its core refers to a person's individuality, distinctiveness, and unique attributes, shaping their thoughts, emotions, and behaviors. Additionally, identity can manifest in collective forms, encompassing group identities like ethnicity, religion, nationality, or gender, where individuals find a sense of belonging. Moreover, identity can extend into virtual realms, with digital avatars enabling individuals to adopt alternative personas or representations. These various identities profoundly influence how individuals perceive and engage with the world, molding their experiences and relationships. Significantly, identity is not rigid or static but fluid and ever-changing, evident through the performative nature exemplified in the aforementioned examples.

In Judith Butler's 1988 work "Performative Acts and Gender Constitution," originally published in *Theatre Journal*, she delves into the phenomenological formation of gender through the repetitive enactment of gendered behaviors. As such, she defines gender as "a constructed identity, a performative accomplishment which the mundane social audience, including the actors themselves, come to believe and perform in the mode of belief" (Butler, p. 520). These actions not only construct gender as identity but also challenge and question the rigidness of any fixed gender construction, highlighting how the creation of a persuasive illusion of gender as a belief object opens up the possibility for contesting its rigid boundaries. In essence, the fluidity of identity can be comprehended by examining gender construction, which provides a framework for exploring identity considerations more broadly.

Central to the reconsideration of fixed identity is the notion of performing or acting. According to Butler, philosophers may not explicitly delve into theatrical acting, but they frequently examine the notion of "acts" in discourse, which shares semantic connections with theories of performance and acting. For instance, the notion of "speech acts" presented by Searle (1979) exemplifies verbal assurances and promises that not only indicate a speaking relationship but also establish a moral bond between speakers, representing a significant element in the domain of analytic philosophy of language. Additionally, the domain of moral philosophy known as "action theory" seeks to understand the nature of human actions before delving into what one ought to do (Frey 2020). Moreover, the phenomenological theory of "acts," championed by Edmund Husserl, Maurice Merleau-Ponty, and George Herbert Mead, among others, aims to elucidate how social agents shape social reality through language, gestures, and various symbolic social signs (Alloa et al. 2019). While phenomenology might seem to assume the existence of a choosing and constituting agent before language (who appears as the sole source of constitutive acts), there is also a more radical application of the doctrine of constitution that regards the social agent as an object rather than the subject of constitutive acts.

In terms of agency, identity is a set of conscious choices made to conform to (or not) a set of predetermined social prescriptions, which unfold over time. Similarly, Simone de Beauvoir's claim that "one is not born, but becomes a woman" reinterprets the doctrine of constituting acts from the phenomenological tradition, highlighting that gender is not a stable identity but rather an identity constituted over time through stylized repetitions of acts (Pettersen 2020). Gender is shaped by the stylization of the body and the mundane enactment of bodily gestures, movements, and behaviors that create the illusion of a fixed gendered self. This understanding shifts gender from being a substantial model of identity to one that requires a conception of a constituted social temporality. In other words, the traditional idea of gender as a fixed and inherent part of who one is shifts to a view where gender is shaped by one's social and cultural context; gender is not something that we are born with, but something that is formed over time through interactions and experiences in society. As such, gender identity can be understood as a performative accomplishment, constructed and reinforced through social sanctions and beliefs. By recognizing that gender is constituted through acts that are internally discontinuous, the potential for gender transformation and contesting reified conceptions of gender identity becomes apparent.

Moreover, the connection between binary genders and the heterosexual contract can be examined in how cultural constructs of sex and gender have become conjoined and reified as natural over time. The social heterosexual contract argues that the association of a natural sex with a discrete gender and a supposedly natural attraction to the opposite sex is a cultural construct serving reproductive interests. Feminist cultural anthropology and kinship studies demonstrate how cultures regulate and guarantee reproduction, enforced by taboos and punishment. With such considerations, phenomenology can assist in understanding the embodied and enacted nature of gender, considering how acts that constitute gender are collective and not solely individual. Gender is seen as an act, a repeated performance with both individual and public dimensions, strategically designed to maintain gender within its binary frame.

Therefore, a gendered identity is not a simple individual choice or a passive inscription on the body by cultural codes. Instead, it is a public and performative act that operates within already existing cultural conventions. The body does not pre-exist these cultural conventions, nor does it simply passively receive them. Rather, actors are always within the terms of the performance, enacting interpretations within the confines of existing directives. Thus, the exploration of gender as a public and performative act reveals that identities, including gender, are not fixed or predetermined but rather negotiable and socially constructed. The concept of gender as a script that can be enacted in diverse ways challenges the notion of a stable and essential identity. This fluid understanding of identity extends beyond gender and applies to other aspects of selfhood as well. The negotiation of identities within the cultural context allows for the possibility of subversion, transformation, and liberation. In the new technocultural realm of avatars, the negotiation of identity takes on new dimensions. Avatars provide a digital embodiment that allows individuals to experiment with various personas and expressions. As virtual representations, avatars enable people to transcend physical limitations and explore identities beyond their everyday lives. The malleability of avatars opens up exciting opportunities for self-discovery and the expression of identity in novel ways.

5.3 Avatars in Cultural Heritage: Living the Past

When approaching avatars from a contemporary cultural perspective, the term often refers to a representation of an individual or a user in the digital realm, typically in virtual worlds, video games, or online platforms, such as in the *Assassins Creed* (2007–2020) or *God of War* (2005–2022) series. These avatars can take various forms, such as humanoid characters, animals, objects, or abstract symbols, and are designed to represent the user's identity, preferences, and actions within the virtual environment. At the same time, the concept of avatars in art history is closely tied to the advent of digital technology and the rise of virtual reality and online communities. With the increasing integration of digital media and art, avatars have become significant elements in digital art, interactive installations, and video games, blurring the lines between the real and virtual worlds.

The rapid growth of virtual experiences has especially paved the way for innovative advances in immersive technologies, providing new and engaging ways to interact with avatars for cultural heritage experiences. Avatars, which can represent individuals, historical figures, guides, or fellow visitors, have emerged as powerful tools in heritage contexts. Their design and implementation offer the potential to create captivating, interactive, and customizable environments that foster a profound connection with cultural heritage. As highlighted by Bekele and Champion (2019), avatars bridge the gap between visitors and historical figures, facilitating direct engagement and personalized interactions. Through the embodiment of these avatars, for example, visitors can step into the roles of specific historical figures, immersing themselves in their lives and experiences. The approach elevates the level of interaction and emotional connection, offering visitors a unique and profound encounter with the past. This immersive and experiential approach to learning enhances understanding of history on the part of the visitor by allowing them to actively participate in historical narratives. By embodying an avatar that represents a particular historical figure, visitors can gain insights into their thoughts, emotions, and motivations; they can witness events through the perspective of the avatar, interact with other characters, and make decisions that shape the outcome of the narrative. The interactive experience offers a deeper level of engagement and fosters a sense of empathy and connection with the past. By immersing themselves in these experiences, visitors can develop a deeper appreciation for the human stories that have shaped our world. The use of these personifications in cultural heritage institutions opens up new avenues for storytelling and interpretation, enabling institutions to go beyond traditional displays and static exhibits, offering dynamic and interactive experiences that have the potential to both captivate and educate visitors. Avatars can bring historical figures to life in a way that transcends time and space, making history more accessible, relatable, and engaging for a wide range of audiences. These historical embodiments have also emerged as interactive guides in the realm of cultural heritage, offering immersive storytelling experiences that enrich visitors' engagement with the past.

Bozzelli et al. (2019), for instance, emphasize the pivotal role of avatars in bringing cultural heritage to life by serving as knowledgeable companions. These virtual guides provide visitors with contextual information, engage them in interactive narratives, and enhance their overall understanding of the heritage site. By interacting with these avatars, visitors can explore historical sites in a guided and informative manner, gaining a deeper appreciation for the significance and historical context of these locations. The use of these embodiments as interactive guides not only enriches visitor experience, but also allows heritage institutions to adapt to diverse learning styles and preferences, creating more inclusive and engaging experiences for all. As interactive companions, avatars offer personalized guidance to visitors, tailoring their explanations and interactions based on individual interests and preferences. Afforded by the latest generation of large language models (LLM) like ChatGPT, avatar guides can tailor responses to the age, language, and preferences of an audience—one explanation can be offered to a tour of school children, while mature visitors would have another, more sophisticated version. This tailored approach allows visitors to ask questions, seek clarification, and delve deeper into specific aspects of the cultural heritage being presented. In as much, they act as virtual interpreters, bridging the gap between visitors and the historical or cultural context they are exploring.

The interactive nature of avatars fosters engagement and active participation from visitors. Through conversations and interactions with the avatar guide, visitors can actively shape their own experience and explore the heritage site from different perspectives. Avatars provide a dynamic and responsive interface, enabling visitors to delve into specific areas of interest, access additional information, or embark on guided tours tailored to their preferences (Viñals et al. 2021). This interactive engagement not only enhances visitors' understanding but also makes the experience more memorable and impactful. Furthermore, avatars have the advantage of adaptability and scalability and can cater to diverse visitor profiles and preferences, providing guidance and content that aligns with individual interests and accessibility needs. In all, these personalized tour guides can also support multiple languages, accommodate different learning styles, and adjust the level of detail or complexity to meet the visitor's requirements. The flexibility allows for a more inclusive and accessible experience, ensuring that a wider range of visitors can engage with and appreciate the cultural heritage being presented.

Beyond merely facilitating tour guide-visitor communication, avatars can also facilitate social interaction and collaboration among and between visitors within physical and virtual spaces. Foster et al. (2022) emphasize the significance of avatars in enabling shared experiences and collective exploration of cultural heritage. In virtual environments, avatars can represent other visitors, creating opportunities for social engagement and community building. Virtual spaces equipped with avatars offer visitors the ability to connect and interact with each other in a virtual community. Through their avatars, visitors can engage in conversations, share insights, and exchange perspectives. This social interaction fosters a sense of community among visitors, creating a collaborative environment where collective knowledge creation and interpretation of cultural heritage can thrive. By connecting with fellow visitors through avatars, individuals have the opportunity to contribute to the overall understanding and interpretation of cultural heritage. Through discussions and shared experiences, visitors can gain new insights, challenge preconceived notions, and develop a richer understanding of the heritage site. The collaborative approach facilitated by avatars promotes active participation and a sense of ownership among visitors, empowering them to actively contribute to the collective knowledge and interpretation of cultural heritage.

Furthermore, the social interaction facilitated by avatars can lead to the formation of communities of interest or practice, where individuals with shared interests or expertise can connect and collaborate. Through these communities, visitors can continue to engage with cultural heritage beyond the museum visit, exchanging knowledge, sharing resources, and deepening their understanding and appreciation of the site. The use of avatars to facilitate social interaction and collaboration aligns with the evolving nature of museums and cultural institutions. It embraces the idea of co-creation and participatory experiences, recognizing the valuable contributions of visitors as active participants in the interpretation and preservation of cultural heritage. Leveraging avatars as social representations can allow museums to create dynamic and engaging virtual spaces that foster social connections, facilitate knowledge exchange, and cultivate a sense of belonging and community among visitors.

The research conducted by Chen and Kent (2020) highlights the transformative potential of avatars in facilitating cultural exchange and understanding within virtual museum projects. By representing individuals from diverse cultural backgrounds, avatars serve as intermediaries that bridge the gap between different cultures, promoting dialogue, empathy, and intercultural interactions. In their study, Chen and Kent explored the use of avatars to facilitate cross-cultural experiences in a virtual museum setting in Second Life. These avatars, representing individuals from different cultural backgrounds, played a pivotal role in creating a platform for visitors to engage in meaningful intercultural interactions. Through interaction with avatars, visitors had the opportunity to gain insights into unfamiliar cultures, challenge stereotypes, and broaden their perspectives. Avatars acted as cultural ambassadors, providing visitors with a unique opportunity to engage in dialogue and exchange ideas with individuals from diverse cultural perspectives. Through these interactions with AI-driven avatars, visitors were able to ask questions, share experiences, and learn about different cultural practices, beliefs, and traditions. The avatars facilitated meaningful conversations and fostered a sense of empathy, helping visitors develop a deeper understanding and appreciation of cultural diversity. By transcending physical and geographical boundaries, avatars enable visitors to overcome the limitations of traditional museum settings and engage in intercultural exchange within virtual heritage spaces. Through the immersive and interactive nature of the virtual environment, visitors can step into the shoes of different cultural avatars, experiencing cultural contexts and perspectives firsthand.

The use of avatars as intermediaries for cultural exchange promotes cultural understanding, breaks down barriers, and fosters mutual respect among visitors. The strategy challenges preconceived notions and biases, allowing individuals to explore and appreciate the richness and complexity of different cultures. Through their non-human presence, avatars facilitate a safe and inclusive space for dialogue, encouraging visitors to ask questions, share personal experiences, and engage in meaningful cross-cultural interactions. Drawing upon the foundations of human psychology and social interactions, these avatars are equipped with animations and response mechanisms that mimic genuine human reactions. The subtleties in their gestures, the modulation in their virtual voices, and their programmed responsiveness to user emotions and inquiries are all geared to evoke a sense of empathy. The example presented by Chen and Kent showcases the potential of avatars in facilitating intercultural exchange and understanding within virtual heritage spaces. The use of avatars as cultural intermediaries opens up new avenues for fostering dialogue, empathy, and appreciation for diverse cultures. It offers a transformative and immersive experience that goes beyond the physical constraints of traditional museum settings, providing visitors with a unique opportunity to explore, connect, and learn about different cultures in a virtual environment.

Another compelling application of avatars in cultural heritage can be seen in the work of Chung (2015). Such research focuses on utilizing avatars to reanimate historical artifacts and objects. By assigning avatars to inanimate objects, such as ancient sculptures or paintings, visitors were able to engage with these artifacts in a dynamic and interactive manner. These engaging characters brought the objects to life, offering historical context, storytelling, and interactive elements, which not only enhanced understanding of the objects but also sparked curiosity and emotional connections. In this way, avatars can act as mediators, breathing life into artifacts and enabling visitors to engage with heritage in novel and captivating ways. Therefore, avatars have emerged as transformative tools in cultural heritage contexts, offering dynamic and interactive encounters that enhance visitors' engagement with the past. By representing historical figures, serving as knowledgeable guides, facilitating intercultural exchange, and reanimating artifacts, avatars bridge the gap between visitors and cultural heritage, fostering a deeper connection and understanding. Their interactive and immersive nature allows visitors to step into different roles, receive contextual information, and collaborate with others in virtual spaces, creating participatory learning experiences. As technology continues to advance and interoperable avatar creation becomes more accessible, avatars hold immense potential for shaping the future of immersive technologies in cultural heritage contexts. Through their capacity to make heritage education more accessible, interactive, and enjoyable for all audiences, avatars are paving the way for innovative and inclusive cultural heritage experiences. As cultural institutions embrace the transformative capabilities of technology, they are poised to create captivating and memorable encounters that redefine the ways we engage with and interpret our shared cultural legacy.

5.4 Avatar Preferences in Cultural Heritage: Demographic and User Engagement

Avatars have become increasingly prominent in the realm of art and culture, particularly in cultural heritage experiences, where they serve as virtual representations that allow users to engage with historical and recreated worlds. These digital embodiments of users provide the ability to understand the world from different perspectives, but personal preferences and individual experience also dictate how the intended cultural experience will be received. Therefore, understanding the preferred avatars for engaging in historical or recreated worlds and scenarios requires a knowledge of the user/player type and personality of the respective individual. Furthermore, the background knowledge and demographics of the individuals are also influential on an engaging and impactful experience. The following case study involving undergraduate and graduate students (some majoring in art history, while others have no background in the field) is useful for those creating avatars and cultural heritage experiences, especially when considering designing for XR. In order to capitalize on the strengths of the new digital medium, including immersion, engagement, and presence, the end user needs to be considered. The heuristic study looked at different experiences, preferences, learning styles, and expectations relating to educational gaming and found that effective game design and development need to consider the target audience to optimize user experience and learning outcomes (Hutson 2022).

Research on the perceptions of students regarding gamification and games in education has been extensive. Malone identified three elements contributing to intrinsic motivation in games—Challenge, Fantasy, and Curiosity (Malone 1981). Rieber et al. (1998) further investigated instructional design's influence on intrinsic motivation and identified four key elements—Control, Challenge, Curiosity, and Contextualization. Building on these studies, Korhonen et al. (2009) categorized Playful Experience (PLEX) in gameplay into twenty types, including captivation, competition, exploration, and sensation, among others. Given the wide range of

Playful Experiences, the enjoyment and engagement of different gamers are influenced by their learner type and demographics. Kim (2013) studied the preferences of engineering students for PLEX categories and found that gender played a crucial role in determining which activities were most enjoyable. Male students preferred challenge, completion, and simulation, while female students favored exploration, relaxation, and Fantasy. These gender-related preferences should be taken into consideration when gamifying learning to meet the specific needs and expectations of the audience.

Different preferences among gamers also give rise to distinct gamer types. Bartle (2010) classified gamer types into eight variations based on a study of Massively Multiplayer Online Games (MMOGs), which are inherently social in nature. Griefer gamers tease or harass other players to gain a reputation, even if negative. Opportunist gamers are explorers who take chances when presented, but they prefer exploration over confronting obstacles. Politician gamers play strategically to achieve long-term goals, exploiting other players to enhance their reputations. Planner gamers are also strategic but set specific goals before acting in a game. Friend gamers prioritize nurturing friendships with other players. Hacker gamers are intuitive and interested in discovering new objects, events, or places. Networker gamers enjoy making new friends but evaluate the benefits of these friendships and may discard them if necessary. Lastly, scientist gamers are interested in understanding cause and effect within the game, engaging in systematic play and conducting experiments. Understanding player types is essential for designing appropriate and engaging games that lead to successful learning outcomes. By assessing player types at the beginning of a class or experience, designers and instructors can assign appropriate roles and tailor the experience to meet the diverse preferences of their audience. This consideration of player types can enhance the overall gamification of learning and foster deeper engagement with art and cultural heritage in immersive XR environments.

As such, the study under discussion sought to gain insights into preferences of students for different types of games and gaming platforms. To do so, the participants were asked whether they played games and what kinds, including video games, board games, card games, party games, or others. The majority of respondents confirmed that they played games, while a minority answered "maybe," and a small minority responded in the negative. Notably, there were no statistically relevant differences in game-playing frequency between male and female students or between art history majors and non-majors. However, when the data was analyzed based on academic level, undergraduate and graduate art history majors exhibited some differences in their responses. All of the undergraduate students claimed to play games, while only 87.10% of graduate students responded the same way. The ages of the respondents also proved interesting, revealing a bell curve pattern mapping gameplay frequency. Game-playing was most prevalent among first-year college students, with 88% of respondents in the 18–24-year-old category confirming they played games. The 25–34-year-old category had the highest game-playing frequency at 91.30%. Both the 35–44-year-old and 45–54-year-old categories had a game-playing frequency of 90%. However, in the 55–64-year-old category, only 66.67% of respondents confirmed playing games, indicating a decline in gameplay frequency with increasing age. Regarding gaming platforms, the preferences of students were as follows: 31.15% preferred mobile gaming on phones or tablets, 24.59% preferred gaming on computers (PC/Mac), 13.11% enjoyed board games, 11.48% preferred PlayStation, 6.56% favored Xbox, 4.92% chose Nintendo Switch, 3.28% opted for VR devices, 3.28% enjoyed party games, and 1.64% were into augmented reality gaming. The top two preferred gaming platforms were mobile gaming and gaming on computers, with the majority of respondents selecting these options.

While there were minimal variances in preferences between art history majors and non-majors, significant differences were observed when the data was separated by gender. Male participants overwhelmingly preferred gaming on computers (PC/Mac), whereas female participants showed a clear preference for mobile gaming on phones and tablets. Similarly, age and gender differences were apparent among undergraduate and graduate art history students. Undergraduates preferred Nintendo Switch as their preferred gaming platform, while graduates leaned more toward mobile gaming. But broken down by age groups, the preferred gaming platforms also changed. The first group (18–24) showed a preference for mobile gaming, the second group (25–34) favored gaming on computers, the third group (35–44) ranked mobile gaming and VR devices as equally popular, the fourth group (45–54) preferred mobile gaming, and the last group (55–64) showed a preference for board games and party games. These findings provide valuable insights into the preferences of different user demographics, guiding the design and development of avatars and cultural heritage experiences that cater to diverse audiences in immersive XR environments.

The study also explored experiences with and preferences for gamified learning. Participants were asked whether they believed that learning material presented as a game would be easier to learn. Among the respondents, 53.97% agreed that gamified learning would be beneficial, while 38.1% responded "maybe," and 7.94% definitively stated that they would not find it helpful. There were no significant differences between male and female responses. However, when considering majors versus non-majors, art history majors showed a greater preference for gamified learning, with 59.46% responding in the affirmative, compared to 44.83% from non-majors. The preferences became more pronounced when differentiating undergraduate and graduate majors. All of the undergraduate art history students (100%) agreed that learning through gamification would be beneficial, while only 51.61% of graduate students said the same. Furthermore, when broken down by age, preferences for gamified learning also varied. In the first group (18–24), 50% preferred gamification, in the second group (25–34) 65.22%, in the third group (35–44) 60%, in the fourth group (45–54) 0% (though 83% answered "maybe"), and lastly, in the fifth group (55–64) 66.67%.

The same preferences were also reflected in responses to whether respondents had played educational video games or applications, such as *Oregon Trail* and *Big Brain Academy*. Among the participants, 80.95% stated that they had played such games, 14.29% stated that they had not, and only 4.76% were unsure. It was observed that more female respondents had exposure to educational gaming compared to males. Moreover, a higher percentage of art history majors had experienced educational gaming compared to non-majors. Generally, the results pointed to a greater preference for exposure to gamified learning among art history majors. Interestingly, even though both graduate and undergraduate populations had the same response when asked if they had played games, significantly more undergraduates found gamification of learning beneficial and preferential. When broken down by age, there was also a difference in how many in each group had played educational games. Among the first group (18–24), 76% answered affirmatively, in the second group (25–34) 91.3%, in the third group (35–44) 80%, in the fourth group (45–54) 66.67%, and lastly, in the fifth group (55–64) 66.67%.

Students were further surveyed on their preferences and experiences with game types, genres, and mechanics. Among the respondents, 46.03% preferred to play games alone, 19.05% preferred cooperative gameplay, and 34.92% had no preference. When considering gender differences, the preferences for gaming platforms carried through into preferences for gameplay styles. Female respondents significantly preferred playing games alone (46.51%) rather than cooperatively (11.63%), whereas male respondents claimed equal preference for playing alone and cooperatively (37.5% each). Due to the proportion of female art history majors, it was not surprising that they showed a greater preference for playing games alone (57.89% in aggregate, and 66.67% for undergraduates and 54.84% for graduates), while non-majors leaned more toward cooperative gameplay (37.93%). However, it should be noted that there was a more even distribution of preferences among non-major respondents. Additionally, preferences shifted when broken down by age. Among the first group (18–24), 36% had no preference, while preferences for playing alone and cooperatively were ranked the same at 32% each. In the second group (25–34), 47.3% preferred playing alone, in the third group (35–44), 70% preferred playing alone, in the fourth group (45–54), 66.67% preferred playing alone, and lastly, in the fifth group (55–64), 66.67% had no preference.

Next, the study examined the elements preferred in gamified learning experiences. Among the respondents, 47.06% selected Curiosity, 33.33% chose Challenge, and 19.61% opted for Fantasy as the most preferred elements. The results showed relatively similar preferences across gender lines, with both males and females ranking Curiosity as their top choice and Challenge as the second-most preferred element. However, males showed a higher preference for Challenge compared to females (35.71% vs. 28.57%), while females expressed a greater interest in Fantasy (22.86% vs. 14.29% among males). Majors and non-majors exhibited differences in their preferences, with 56.67% of art history majors preferring Curiosity, while non-majors leaned toward Challenge (41.67%). Notably, there was a more even distribution among preferences among non-majors.

Moreover, both graduate and undergraduate students ranked Curiosity as their top choice, but among undergraduates, Curiosity was tied with Fantasy, with no interest in Challenge. On the other hand, graduate students' preferences were ranked as Curiosity first, followed by Challenge, and then Fantasy. When broken down by age, the preferences for game play elements were as follows: Curiosity was preferred by the first group (18–24), Curiosity was the top choice for the second group (25–34), Challenge and Curiosity were equally preferred for the third group (35–44), Curiosity was vastly preferred for the fourth group (45–54), and Challenge was preferred for the fifth group (55–64).

The survey of students on avatar preferences and game genres for cultural heritage experiences revealed valuable insights (Table 5.1). Casual Games emerged as the most preferred genre, particularly among female students and non-majors, suggesting the significance of incorporating casual elements in engaging experiences. Casual Games, like Candy Crush, by definition, emphasize gameplay that is straightforward and easily accessible. They are characterized by their simplicity in mechanics and rules, making them easy to pick up and play without significant time investment. Action-Adventure and Role-Playing Games were popular choices, with distinct preferences between genders. Integrating elements from both genres could cater to diverse user groups. Simulation Games and Action Games also showed appeal, highlighting opportunities for interactive historical exploration. Party Games and Digitized Board/ Card Games had limited popularity, necessitating alternative approaches for fostering social interaction. Understanding these preferences can guide the design of culturally engaging experiences, creating inclusive journeys through cultural history.

Following up on the preferred genres, students were asked what types of settings they most preferred for their games and ranked the following in order of preference (Table 5.2). The analysis of the results by gender showed that Mystery Stories were the favorite setting for both males and females. Females ranked Suspense and Fantasy equally as their second preference, while males preferred Fantasy and Science Fiction as their second and third choices, respectively. Major and non-major students also shared a preference for Mystery Stories as their top choice. Additionally, undergraduate majors favored Mystery, followed by Suspense and Fantasy, while graduate majors preferred Mystery, followed by stories set in the past and Science Fiction. The preferences for game settings also varied by age group, with the younger age

Table 5.1 Preferred game genres by student

Game genres	Overall	Male	Female	Art history majors	Non-majors	Undergraduate majors	Graduate students
Casual games	1	2	1	1	2	2 (tie)	1
Action-adventure	2	1	2	2 (tie)	1	1 (tie)	2 (tie)
Role-playing games	3	3	3	3	3	3	3
Simulation games	4	4 (tie)	5 (tie)	4	5	1 (tie)	4
Action games	5	1 (tie)	5 (tie)	5	1 (tie)	3 (tie)	5
Party games	6	–	–	–	–	–	–
Digitized board/ card	7	–	–	–	–	–	–
Strategy games	8	–	–	–	–	–	–

Table 5.2 Preferred game settings

Rank	Preferred game settings
1	Mystery stories
2	Fantasy stories
3	Set in the past
4	Science fiction stories
5	Suspense stories
6	Romance
7	Make you nostalgic for your youth
8	Set in the modern day
9	Post-apocalyptic stories
10	Set in the far future
11	Family
12	Set in the near future
13	Horror stories

group (18–24) ranking Mystery, Fantasy, and Suspense as their top choices, while the older age group (55–64) ranked Modern Day, Mystery, and Romance as their preferred settings.

Following up on the preferred characters, students were asked what types and characteristics they most preferred for their games and ranked the following in order of preference for (Table 5.3). The analysis of the results by gender revealed some interesting preferences. Males showed a preference for games played in the first-person point of view (no visible character), while females preferred the ability to customize their player characters. Both genders ranked visually realistic and visually stylized characters as their second and third preferences, respectively. Additionally, art history majors also shared a preference with females for the ability to customize their player characters. On the other hand, non-majors favored visually stylized characters as their second choice. Among undergraduate majors, the top choice was playing as their own race/ethnicity, which tied with the ability to customize characters. Graduate majors ranked customizing characters as their first choice, followed by visually realistic characters. When considering preferences by age, the results varied. The youngest group (18–24) preferred customizable characters as their top choice, followed by visually realistic characters. The second age group (25–34) preferred visually realistic characters, followed by customizable characters. The third group (35–44) equally preferred characters played in the first-person point of view and visually realistic characters. Both the fourth (45–54) and fifth (55–64) age groups favored characters played in the first-person point of view, followed by customizable characters. Overall, the preferences for game characters were diverse, with players valuing the ability to customize their characters and have visually realistic or stylized appearances. The first-person point of view was more preferred by males and older age groups. These findings indicate that players' preferences for characters are

Table 5.3 Preferred character types and characteristics

Rank	Preferred characters
1	Customized player characters
2	Visually realistic characters
3	Visually stylized characters
4	Characters played in the first person
5	Characters that reflect player's gender
6	Characters that reflect player's age
7	Non-human player characters
8	Characters that reflect player's race
9	Character that is younger than player

influenced by various factors, including gender, major, and age. Understanding these preferences is crucial for game designers and developers when creating inclusive and engaging gaming experiences that cater to different player types and demographics.

Next, students were surveyed on their motivations relating to game play and association with their gamer type (Table 5.4). The analysis of the results indicates that both genders, as well as majors and non-majors, prioritize Captivation, Discovery, and Challenge as their top-three motivations for gameplay. These motivations resonate across different demographics, suggesting that the desire to be captivated, to explore new elements, and to be challenged are fundamental aspects of the gaming experience that appeal to a broad audience. However, when comparing art history majors to non-majors, there are some variations. Art history majors place a higher preference on Discovery, ranking it as their top motivation, followed by Captivation and Exploration. On the other hand, non-majors ranked Captivation as their top motivation, followed by Challenge and Discovery. This finding may be attributed to art history majors' academic background and inclination toward exploration and discovery. When examining the motivations by age group, some interesting patterns emerge. The youngest age group (18–24) prioritized Captivation, Discovery, and Challenge. The second age group (25–34) had a similar preference, with Captivation, Discovery, and Exploration as their top three motivations. The third age group (35–44) ranked Exploration, Discovery, and Captivation as their top motivations. The fourth age group (45–54) valued Challenge, Captivation, and Discovery the most, while the fifth age group (55–64) prioritized Challenge, Discovery, and Fellowship. Thus, the motivations for gameplay indicate that players are primarily driven by the desire for captivating experiences, opportunities for exploration, and the thrill of facing challenges. These motivations transcend gender and academic background, indicating that game designers should focus on creating captivating and challenging gaming experiences that cater to the sense of discovery and exploration desired by players across age groups and demographics.

Finally, with regard to gamer types, students were asked to self-select their own and the preferences were ranked (Table 5.5). The analysis of the results reveals interesting insights into self-identified gamer types among the students surveyed.

Table 5.4 Motivation for gameplay

Rank	Motivations
1	Captivation
2	Discovery
3	Exploration
4	Challenge
5	Completion
6	Fantasy
7	Competition
8	Expression
9	Relaxation
10	Control
11	Fellowship
12	Simulation
13	Thrill
14	Sensation
15	Nurture
16	Sympathy
17	Eroticism
18	Subversion
19	Sadism
20	Suffering

Both genders, male and female, ranked Opportunist as their most self-identified gamer type, with almost identical scores. However, there were significant differences between male and female preferences for other gamer types. Males tended to rank Politician, Planner, Networker, Griefer, and Scientist as their next preferred types, while females ranked Friend, Hacker, and then Planner and Scientist tied for fourth, with Politician and Networker tied for fifth. Furthermore, the preferences for gamer types differed between art history majors and non-majors. While both major and non-major students ranked Opportunist as their top gamer type, majors had a more balanced distribution of preferences, including Hacker, Politician, Planner, Scientist, Friend, and Networker, which were all ranked at different levels. On the other hand, non-majors had a more concentrated preference for Friend, Networker, and Griefer as their top choices, while Planner and Scientist were tied for last. Additionally, the preferences for gamer types varied among undergraduate and graduate majors. Undergraduates identified a broader range of types, including Opportunist, Friend, Hacker, and Planner. Graduate majors had Opportunist as the most common gamer type, followed by Politician and Hacker tied for second, then a tie for third between Planner and Scientist, and finally, Friend and Networker. Lastly, when examining the self-identified gamer types by age group, some interesting patterns emerged. The youngest age group (18–24) ranked Friendship type as their first choice, followed

Table 5.5 Self-identified gamer types

Rank	Gamer types
1	Opportunist
2	Friend
3	Hacker/politician/planner
4	Scientist/networker
5	Griefer

by Opportunist and Networker. The second age group (25–34) ranked Opportunist first, followed by Politician, Hacker, and Scientist, all tied for second. The third age group (35–44) had Opportunist, Planner, and Hacker ranked first equally, followed by Politician and Scientist, which were also tied in the second position. The fourth age group (45–54) ranked Opportunist first, followed by a tie between Planner and Hacker. Lastly, the fifth age group (55–64) ranked Politician, Friend, and Scientist as their top choices. Thus, the survey indicates that both genders tend to identify as Opportunist gamers, with variations in preferences for other gamer types. The preferences for gamer types also vary between art history majors and non-majors, as well as between undergraduate and graduate students, and among different age groups. Understanding these differences can provide valuable insights for game designers and educators to tailor gaming experiences to the preferences and self-identified gamer types of their target audience.

The survey concluded with two open-ended questions, inviting students to express their preferences for art history games and provide additional considerations for game developers. In response to the question about their preferred art history game, students mentioned a strong interest in antiquity, particularly ancient Greece and Rome, as well as classical gods. Other popular choices included Egypt, Tang China, Sengoku Japan, Medieval, Renaissance, and modern settings like nineteenth- and twentieth-century styles. Some students expressed a desire for an interactive platform where they could explore different artist styles and mediums from various eras to gain a deeper understanding of art history.

The responses to the open-ended question aligned with the preferences identified in the survey, indicating a clear preference for games set in the distant past rather than the recent past. The second open-response question asked students for other considerations that developers should keep in mind when creating educational art history games. The responses highlighted a desire for engaging, immersive, and historically accurate experiences. Students emphasized the importance of presenting factual information in the game, possibly through story narratives or collectibles like notes found in the virtual world. Historical accuracy was seen as a crucial element for creating an authentic and educational experience.

Another noteworthy point that emerged from the responses was the tension between creating an educational experience and a fun gaming experience. Some students expressed concerns that labeling a game as "educational" might lead to it being perceived as less of a game and more of an interactive learning tool. Balancing the educational aspect while still providing an entertaining and enjoyable gaming experience was seen as a challenge.

To gain further insights, the responses were analyzed based on different demographics, such as major, gender, and age groups. Among the different age groups, preferences for game settings and elements varied. The youngest age group (18–24) showed a preference for games set in antiquity with a focus on realistic elements. The second age group (25–34) preferred settings in the nineteenth- and twentieth centuries, expressing some skepticism about the survey instrument and confirming a preference for puzzles or playing as famous artists. The third age group (35–44) favored ancient and non-western settings with an emphasis on sensory immersion and historical accuracy. The fourth age group (45–54) leaned toward medieval and Renaissance settings, particularly for games set in VR and incorporating job skills. Finally, the fifth age group (55–64) favored a twentieth-century American setting but was insistent that a gaming element should not be mandatory for an art history program.

Recommendations for designing, developing, and deploying avatars for cultural heritage experiences can be tailored based on the specific demographics of the target audience. For non-art history majors, who are primarily taking classes for General Education requirements, it is essential to consider their motivations and perceived relevance of the material presented. They are less likely to prefer gamified learning experiences and are more interested in playing alone or cooperatively. Designers should aim for wider interests when incorporating game elements, and Action-Adventure should be the preferred genre. Mystery Stories are their preferred settings, and visually stylized characters are more appealing to this population. Their motivations are primarily driven by Captivation, Challenge, and Discovery, and they identify as Opportunist gamers. On the other hand, when designing for art history majors, who are already motivated by the subject matter, the focus should be on crafting meaningful and engaging activities and outcomes. Art history majors prefer gamified learning experiences and have more exposure to educational games. They prefer playing alone and are most motivated by Curiosity. Casual Games and Mystery Stories are their preferred genres. They lean toward playing as customizable characters and are primarily motivated by Discovery, Captivation, and Exploration. Their most common gamer type is Opportunist.

When designing for undergraduate art history majors, it is crucial to acknowledge their interest in gamified learning, as they find it beneficial and preferential. They prefer playing alone and are driven by Curiosity and Fantasy. Their preferred genres are Simulation and Action-Adventure, and Mystery is their top choice for settings. They prefer playing as characters of their own race/ethnicity but also appreciate the ability to customize characters. Their motivations in gameplay are Discovery, Exploration, and Fantasy, and they identify as Opportunist, Friend, Hacker, and Planner gamers. For graduate art history majors, who are typically older and have more life experience, designers should consider their preference for mobile gaming. While they may not be as interested in gamified learning, they still prefer playing alone and are motivated by Curiosity. Their preferred genres are Casual Games, Action-Adventure, and Role-Playing Games. Mystery and stories set in the past are their top choices for settings. Graduate majors also prefer playing as customizable characters and visually realistic characters. Their motivations in gameplay are Discovery, Captivation, and Exploration, and they identify as Opportunist, Politician, Hacker, Planner, Friend, and Networker gamers.

It should be recognized that preferences and behaviors change over time. As shown by the bell curve mapping gameplay across different age groups, designers should consider the longitudinal differences. Combining this with the preferences of each generation for platforms, game genres, and player types will lead to more successful outcomes in designing avatars for cultural heritage experiences. Understanding the specific demographics and preferences of the target audience is key to creating engaging and meaningful experiences in the realm of art and cultural heritage. When cultural heritage institutions are designing experiences for the general public, they can leverage the insights gained from the survey results to create more engaging and appealing avatar-based experiences. Understanding the preferences and motivations of different demographic groups allows institutions to tailor their content and presentation to meet the diverse interests of their visitors.

For the general public, which includes both art history majors and non-majors, cultural heritage institutions can focus on providing a wide range of avatars and gameplay options. Since non-majors may be less inclined toward gamified learning experiences, institutions can offer both traditional and interactive approaches to cater to a broader audience. By providing visually stylized characters and a mix of settings that include both antiquity and modern styles, institutions can captivate visitors with diverse interests. Considering the popularity of Mystery Stories among both genders and various age groups, cultural heritage institutions can design avatar-based experiences that incorporate immersive storytelling and interactive elements to engage visitors in historical Mysteries and challenges. This approach can appeal to the general public's curiosity and sense of discovery. Furthermore, incorporating elements that reflect the player's gender, age, and race/ethnicity can enhance the personalization and relatability of the avatar experience. Institutions can provide options for visitors to customize their characters, making them feel more connected to the content and the historical context presented.

Designing avatar-based experiences that offer opportunities for both individual exploration and cooperative gameplay can cater to the preferences of different age groups and gaming styles. For example, while undergraduates may prefer exploring historical settings individually, older age groups might enjoy collaborative experiences with family or friends. To cater to audiences with a background and interest in the arts, who are more interested in mobile gaming and visually realistic characters, institutions can offer experiences that blend historical accuracy with modern technology, such as augmented reality or virtual reality. This approach can create an immersive and visually captivating environment that resonates with this particular audience. Considering the motivational factors identified in the survey, cultural heritage institutions can infuse their avatar-based experiences with elements of captivation, challenge, and discovery. By providing visitors with captivating narratives and challenging puzzles, institutions can create an environment where visitors actively engage with the content and deepen their understanding of cultural heritage. As such, the results of the study provide valuable insights for cultural heritage institutions to create dynamic, personalized, and immersive avatar-based experiences that resonate with the general public. By taking into account the preferences, motivations, and gaming styles of different demographic groups, institutions can design culturally enriching and entertaining experiences that appeal to a broad audience and foster a deeper appreciation for art and cultural heritage.

5.5 Avatars in the Sensory Realm: Multisensory Experiences and Avatars

Taking into consideration the various factors, motivations, and preferences for a given user, incorporating scenario and avatar preferences for a desired demographic and population allows for more multisensory immersion than ever before in simulated environments. The rapid evolution of immersive reality technologies has opened up new possibilities for cultural heritage experiences. From virtual environments with limited realism to the latest advancements that enable multisensory immersion, avatars have become powerful conduits for transporting users back to historical places and times. The latest cutting-edge technologies can enhance the sensory richness of avatar-based encounters, revolutionizing the way we engage with cultural heritage.

Historically, virtual environments were limited to exteriors of ancient cities, lacking realistic texturing and sensory elements. However, with the integration of modern technologies, historical interiors now come to life with digitally reconstructed spaces, complete with furnishings, textures, and architectural details. Through hand-held scanners and Matterport systems, high-resolution digital twins of objects and environments can be created with ease and precision, providing a more authentic and sensorially superior experience compared to traditional descriptions.

Beyond visuals, new technologies are pushing the boundaries of immersion even further. Haptic feedback allows users to interact with visual information through touch, creating a distinct and realistic user experience. Devices like VR controllers and finger-augmentation devices like Senseglove (https://www.senseglove.com/) provide vibrotactile feedback, enhancing performance and accessibility for users, while also allowing for the decoupling of touch and tactile sensation points, creating unique interactions in virtual environments. Moreover, the integration of "smell-o-vision" and olfactory VR introduces historical scents, adding a new dimension to cultural heritage experiences. Visitors can now encounter the aroma of ancient marketplaces, incense in temples, and the ambiance of historical settings, fostering a deeper emotional connection to the past.

One notable company in this realm is OVR Technology (https://ovrtechnology.com/), which has been at the forefront of integrating olfaction into the virtual realm, providing an enhanced sensory experience that bridges the digital and the tangible. Their Virtual Olfactory Exhibit stands as a pioneering venture into a metaverse where the sense of smell takes center stage. In collaboration with various companies and design entities, OVR Technology launched the metaverse exhibit titled "Living with Scents" showcased at the Museum of Craft and Design. This exhibit did not merely feature scented products in the traditional sense. Instead, it explored innovative interfaces through which scents could be delivered, emphasizing creative, artful outcomes. The VR scent hardware developed by OVR enabled participants to virtually "smell" the items presented in this exhibition, providing an immersive experience where users could interact with and understand scents in a novel manner. This innovative approach reflects the potential of virtual reality to transcend visual and auditory boundaries, delving into the profound sensory realm of olfaction, and offering users an enriched, multisensory experience.

As these technologies continue to evolve, the potential for even more immersive cultural heritage experiences expands. Avatars, equipped with expressive capabilities, enable dynamic interactions and storytelling, further enriching encounters with history. By considering demographics and preferences, cultural heritage institutions can tailor avatar-based experiences for different populations, ensuring relevance and engagement. In all, avatars in cultural heritage have moved beyond static visual depictions to become conduits of multisensory immersion. Technological advancements have facilitated historical accuracy, tactile experiences, realistic scents, and expressive interactions, making the virtual journey back in time more authentic and captivating. As we explore the potential of these technologies further, the realm of cultural heritage experiences is bound to witness ever more engaging and transformative encounters, connecting people of all ages and backgrounds to the richness of human history.

References

Abdallah M, Moustafa M, Morsi EM, Ali GR (2022) Archaeometric study and conservation of a goddess Bastet statue from the late period of ancient Egypt. Int J Conserv Sci 13(2):491–514

Alloa E, Chouraqui F, Kaushik R (eds) (2019) Merleau-Ponty and contemporary philosophy. Suny Press

Bartle RA (2010) From MUDs to MMORPGs: the history of virtual worlds. In: International handbook of internet research, pp 23–39

Bekele MK, Champion E (2019) A comparison of immersive realities and interaction methods: cultural learning in virtual heritage. Front Robot AI 6:91

Benito JRL, González EA (2019) Enterprise augmented reality projects: build real-world, large-scale AR solutions for various industries. Packt Publishing Ltd.

Bozzelli G, Raia A, Ricciardi S, De Nino M, Barile N, Perrella M et al (2019) An integrated VR/AR framework for user-centric interactive experience of cultural heritage: the ArkaeVision project. Digit Appl Archaeol Cult Herit 15:e00124

Butler J (2020) Performative acts and gender constitution: an essay in phenomenology and feminist theory. In: Feminist theory reader. Routledge, pp 353–361

Chen JC, Kent S (2020) Task engagement, learner motivation and avatar identities of struggling English language learners in the 3D virtual world. System 88:102168

Chung HJ (2015) The reanimation of the digital (un)dead, or how to regenerate bodies in digital cinema. Vis Stud 30(1):54–67

Dale R (2021) The myths and realities of the Viking Berserkr. Routledge

Foster JK, McLelland MA, Wallace LK (2022) Brand avatars: impact of social interaction on consumer–brand relationships. J Res Interact Mark 16(2):237–258
Frey JA (2020) Revisiting modern moral philosophy. R Inst Philos Suppl 87:61–83
Hutson JL Jr (2022) Gamification of education and learning: heuristic elements, player types, and learning outcomes for art history games
Kim S (2013) Analysis of engineering students' needs for gamification based on PLEX model. J Knowl Data Eng 1(1):1–7
Kirby D, Özkula SM (2023) The dead pig's photo album: affective visual rituals in collective identity formation. Int J Cult Stud 26(5):567–587
Korhonen H, Montola M, Arrasvuori J (2009) Understanding playful user experience through digital games. In: International conference on designing pleasurable products and interfaces, vol 2009, Oct 2009, pp 13–16
Malone T (1981) What makes computer games fun? In: Proceedings of the joint conference on easier and more productive use of computer systems. (Part-II): human interface and the user interface, vol 1981, May 1981, p 143
Masco J (1995) "It is a strict law that bids us dance": cosmologies, colonialism, death, and ritual authority in the Kwakwaka'wakw Potlatch, 1849 to 1922. Comp Stud Soc Hist 37(1):41–75
Monga C, Das AK (2021) Reconstruction of vanishing indigenous cultural threads of Naamghar in Assam. In: Design for tomorrow—volume 1: proceedings of ICoRD 2021. Springer Singapore, pp 793–806
Nizam DNM, Rudiyansah DN, Tuah NM, Sani ZHA, Sungkaew K (2022) Avatar design types and user engagement in digital educational games during evaluation phase. Int J Electr Comput Eng 12(6):6449
Nyord R (2020) Seeing perfection: ancient Egyptian images beyond representation. Cambridge University Press
Octobre S (2020) Youth technoculture: from aesthetics to politics. Brill
Pettersen T (2020) "On ne naît pas femme: on le devient…": the life of a sentence, by Bonnie Mann and Martina Ferrari, eds. Simone Beauvoir Stud 30(2):383–391
Rieber LP, Smith L, Noah D (1998) The value of serious play. Educ Technol 38(6):29–37
Rosen S (2023) Forms of Krishna: collected essays on Vaishnava Murtis. Rowman & Littlefield
Searle JR (1979) Expression and meaning: studies in the theory of speech acts. Cambridge University Press
Tateyama H (2016) Ritual of superiority: Tolai Tubuan performance at the national mask festival in Papua New Guinea. J Soc Océan 21–36
Viñals MJ, Gilabert-Sansalvador L, Sanasaryan A, Teruel-Serrano MD, Darés M (2021) Online synchronous model of interpretive sustainable guiding in heritage sites: the avatar tourist visit. Sustainability 13(13):7179
Walker G, Weidenbenner JV (2019) Social and emotional learning in the age of virtual play: technology, empathy, and learning. J Res Innov Teach Learn 12(2):116–132

Chapter 6
From Simulacra to Reanimation: Resurrecting the (Un)Dead

Abstract The chapter explores the intersection of digital technology and historical reanimation, focusing on how virtual environments and digital tools are used to bring historical figures and narratives back to life. This chapter critically examines the concept of digital necromancy—using technology to "resurrect" historical figures and events for educational and cultural enrichment. It discusses the ethical implications and technological challenges of digitally reconstructing past realities, including the accuracy and authenticity of such representations. The chapter delves into various case studies where digital tools have been employed to recreate significant historical moments and figures, allowing contemporary audiences to interact with and experience the past in immersive ways. It also addresses the potential of these technologies to shape public memory and historical understanding, questioning the boundaries between reality and representation. By analyzing the impact of these reanimations on cultural heritage and collective memory, the chapter provides insights into the transformative effects of digital technologies on our relationship with history. It ultimately argues that while digital reanimation opens up new avenues for education and engagement, it also necessitates careful consideration of the implications for historical authenticity and ethical representation.

6.1 Digital Necromancy: The Historian's Dilemma

In the realm of art and culture, the past comes alive through virtual recreations, resurrecting historical figures, reconstituting structures, and facilitating interactions. Past examples include using avatars to reanimate historical artifacts and objects. By assigning avatars to inanimate objects, such as ancient sculptures or paintings, visitors were able to engage with these artifacts in a dynamic and interactive manner, bringing them to life and adding interactive elements. However, these were animated objects and not people. While the public has become accustomed to animatronics of historical persons, such as Great Moments with Mr. Lincoln, debuting at the 1964 New York World's Fair before finding a home at Disney Land, these performances are neither immersive nor interactive. The profound potential of the latest generation

J. Hutson, *Art and Culture in the Multiverse of Metaverses*, Springer Series on Cultural Computing, https://doi.org/10.1007/978-3-031-66320-8_6

of immersive reality and generative Artificial Intelligence (AI) has opened doors to reviving the (un)dead, challenging the way we perceive and engage with history. Now, historical figures can not only exist in the same physical or virtual space with audiences but can be engaged in conversation with the latest large language models (LLM).

Exploring the ever-expanding theme of digital necromancy, the lens of "Mourning the Death of Art History" by Davis (1994) shall be used to illuminate the intricate complexities of connecting with the past through these transformative technologies. The provocative work adds depth to the contemplation of XR and generative AI, as we delve into the role of digital twins in bringing history to life. The glorification of certain "Golden Ages" and their relevance to the present prompts critical questions about our authentic connection to history. For instance, is it actually possible to have a genuine experience or understanding of the past, or does it merely reflect our contemporary expectations and cultural values? What material, data, and information are required to recreate and resurrect distant historical figures and civilizations? And, finally, if possible, then how "authentic" do we want said experiences to be and for what types of audiences? Within the realm of digital reconstructions and AI applications in art and culture, answers to these questions revolve around the limitations of historical records, the risks of simulacra, biases in existing data, and the lack of perspectives of marginalized groups.

Regardless of the historical accuracy or ethical considerations, the ability to resurrect the dead is not a distant prospect, but a current reality in the realm of digital necromancy, and its prevalence is expected to increase significantly in practice. The boundaries between life and death are being redefined in unprecedented ways with the advent of technology, facilitating the creation of digital twins and AI clones of both the living and the deceased (Fuller et al. 2020). This growing field, commonly known as "digital necromancy" or "digital cloning" has captivated the imagination of scholars and the public alike, turning once-fantastical possibilities into tangible realities (Morse 2024). In this context, "digital necromancy" denotes the use of advanced technologies, such as AI and robotics, to reconstruct or simulate the presence of individuals who have passed away, enabling posthumous interactions with their digital personas. This cutting-edge practice involves gathering data from diverse sources, including online profiles, voice recordings, images, and other digital artifacts, to construct virtual versions of individuals, allowing them to persist in digital form. The convergence of technology and commemoration in digital necromancy carries profound implications, prompting ethical contemplations, unveiling technological frontiers, and presenting intriguing ethical dilemmas. Moreover, as the technology advances, the presence of digital versions of deceased individuals is likely to become increasingly prevalent in our lives.

Recent technological advancements have showcased the extraordinary potential of recreating and engaging with deceased individuals. One remarkable instance of this progress is the creation of Ameca, an AI-powered avatar that embodies the personality of a specific individual (https://www.engineeredarts.co.uk/robot/ameca/). By leveraging large language models (LLM) like GPT-3 and GPT-4, Ameca is capable of engaging in conversational interactions, displaying a wide array of emotions, and even responding to real-time events. Its ability to replicate facial expressions and provide nuanced responses engenders an uncanny sense of presence, blurring the boundary between the living and the deceased. Such groundbreaking developments exemplify how AI and robotics are ushering in a new era of commemoration and interaction with those who have passed away, prompting thought-provoking inquiries into the essence of identity, consciousness, and the ethical implications of resurrecting the dead (Xygkou et al. 2023). As these technologies continue to evolve, the prevalence of digital interactions with the departed is likely to grow, reshaping our experiences with those who are no longer with us.

The longing to commemorate and interact with departed individuals has a deep-rooted history, stretching back to at least the Neolithic era (Rodanés-Vicente et al. 2023). In recent times, remarkable technological progress has brought us closer than ever to fulfilling this age-old aspiration. Advancements in digital twins and AI clones have gone beyond the realm of fiction, enabling conversations with videos of celebrities and historical figures (Hutson and Ratican 2023). Companies like StoryFile (https://storyfile.com/) have pioneered conversational video AI, allowing users to engage with recorded interviews of iconic personalities like William Shatner. Through these interactive experiences, a glimpse into the past is offered, where users can pose questions and receive personalized responses, blurring the lines between the living and the departed. These developments showcase the growing convergence of AI, virtual reality, and human–machine interaction, providing insights into the possibilities and ethical considerations surrounding virtual afterlives. As technology continues to evolve, the prospect of engaging with the deceased through digital means becomes ever more plausible and raises intriguing questions about the nature of existence and the boundaries of human–machine relationships.

The rise of digital necromancy, with its capacity to reconstruct individuals based on their digital footprints, presents compelling ethical dilemmas. A significant concern revolves around the potential for identity theft and the unauthorized recreation of individuals without their explicit consent. As technology advances, it becomes increasingly feasible to generate replicas of people using their online presence, voicemails, or even selfies, blurring the line between consent and exploitation (Kaur et al. 2023). This raises fundamental questions regarding ownership and control over digital identity, as well as the right to determine how one is represented in the virtual realm. At the same time, the advent of generative technologies, AI clones, and digital twins has opened up new frontiers in the realms of AI and robotics. These advancements offer intriguing possibilities for commemorating the deceased, engaging with historical figures, and even shaping interpersonal relationships. However, they also give rise to profound ethical considerations surrounding consent, identity integrity, and the potential distortion of personal narratives. Nevertheless, most research on digital necromancy has thus far focused on utilizing digital media to create clones of figures from the recent past, benefiting from abundant datasets and media, including audio and video recordings of the individuals, such as speeches of famous leaders, along with the rich data available from social media platforms. However, as one seeks to delve further back into history, the available data becomes increasingly scarce, casting doubt on the veracity and accuracy of the resulting digital personas.

6.2 Art History as a Pageant of Corpses Revivified

In her article, "Mourning the Death of Art History" (1994), Davis offers crucial insights into the complexities faced by art historians in their pursuit of understanding and interpreting the past, where she deemed "Art History as a Pageant of Corpses Revivified." In exploring the contrasting perspectives of the artist-biographer Vasari and art historian and archeologist Winckelmann, Davis illustrates how each confronted the challenges of reconstructing the achievements of ancient art in their respective historical contexts. For Winckelmann, the relationship between the viewer and the object of historical interest was far from straightforward; he saw himself as a witness to the departure of both the ancient tradition and its Renaissance revival, leaving the historian grappling with the losses inherent in this process. Restoring the historical context of the "object" of study could lead to its alienation from the present, resulting in its irrevocable loss. Thus, the historian's investigation into the nature of the object would reveal its otherness, belonging to a lost time and speaking to others in a way that contemporary historians could never fully grasp.

These dilemmas and contradictions are evident in the attempt of Winckelmann in particular to reconcile his personal desires, especially his homoerotic fascination with the Greek male nude statues, with his scholarly investigations of their historical significance as epitomes of Greek culture. In such an instance, the position of the historian becomes entangled in questions of gender relations and the distinctions between subjects and objects. The complexities and ambiguities underscored by Davis highlight the challenges faced by historians in navigating the boundaries, laws, and social organization of modern society. Moreover, in the realm of digital necromancy, art historians encounter similar dilemmas as they seek to bring the dead back to life through advanced technologies. Just as Winckelmann grappled with the loss and alienation of historical objects, digital necromancy raises questions about the authenticity of interactions with resurrected individuals and the implications of reviving the departed in the present. The restorative act of the historian in an attempt to understand the past could evoke a sense of otherness, challenging the very nature of identity and historical representation.

Importantly, as Winckelmann (1850) delved into the systematic project of understanding the circumstances that gave rise to Greek art in his *History of Ancient Art*, first published in Dresden in 1764, he found himself confronted with the inherent limitations of historical representation. The Greek ideal, once glorified as an epitome of beauty and achievement, became relativized within the context of history, forever beyond his own reach. The historian possessed only copies or representations of these ideals, which served as catalysts for an insatiable desire to grasp the elusive realities of beauty they symbolized. However, this pursuit remained unending, as the deceased objects could never truly be brought back to life, and the beauty, whether found in works of art or living subjects, always left something more to be desired. Often regarded as the first true "history of art," the treatment elevated art history from a mere chronicle of artists' lives to a higher level, incorporating systematic stylistic analysis, historical contextualization, and iconographical analysis. The work provided essential tools for general criticism, offering detailed descriptions of individual artworks for aesthetic evaluation or historical analysis. As well, the groundbreaking contribution integrated the twin methods of "formalism" and "historicism," exploring the forms of Greco-Roman art, such as the Belvedere Torso, Apollo, and Laocoön, and their historical explanations, including factors like climate. The construct proved instrumental in shaping the development of art history as a scholarly discipline. Yet, despite the profound impact on the discipline, the seminal work is accompanied by the unintentional entanglement of the historian in the subject-object dilemma.

In a manuscript, Winckelmann alludes to a "copy" of an artwork without providing clear details about the original, leaving us to infer that it signifies a trace of a lost object primarily due to its inherent beauty. This intriguing aspect of his analysis opens up a realm of inquiry into the intricate relationship between beauty, imitation, and the process of discovering or replicating Classical art. By emphasizing the significance of beauty in the "copy," Winckelmann implicitly challenges the idea of imitation as merely copying, suggesting that true imitation involves a profound understanding and rediscovery of Classical art's essence. The metaphor is further elucidated, bringing forth an intriguing paradox in the context of what the historian witnesses. According to his metaphor, the observation of the decline and subsequent revival of Greek art does not manifest as a clear-cut loss or downfall, and instead, as with the journey home of the *Odyssey*, is likened to a perpetually setting sun, never fully descending or rising again, and to the endless return of Odysseus, never quite reaching home. This perpetual "downfall" lacks a complete presence or absolute absence, embodying a state of "greatness" without "force." The depiction culminates in a powerful image that serves as the foundation of the discipline of art history, where he finds himself "compelled to believe" that the beautiful object before him is merely a "copy" of a lost original, alluding to the destruction of his "native land." Winckelmann compares this sensation to that of a maiden gazing tearfully at her departing lover, convinced that the distant sail holds the image of her beloved. In this intricate metaphor, Winckelmann acknowledges that art historians, like the loving maiden, are left with nothing but shadowy outlines of their desired objects. Yet, this very indistinctness intensifies their longing for what is lost, driving them to study the copies of the originals with even greater attentiveness than they would have studied the originals themselves if they were in full possession of them. This captivating metaphor not only encapsulates the essence of the discipline but also reveals the profound yearning and dedication art historians possess to unveil the secrets of the past and understand the treasures they study.

Therefore, the Winckelmannian approach to art history, as illustrated in his practice, reveals a profound interweaving of departure and witnessing—a poignant act of mourning for the loss of the history of art. Art history, in essence, embodies the continual observation and acknowledgment of what has been lost, bringing forth the life-in-mourning aspect of the discipline. To fully embrace the history of art as history, art historians must confront the irreparable loss of the objects they study. This confrontation, however, raises a dilemma: Should art history be a perpetual attempt to bring back to life the departed objects, revivifying them with their presumed original ideas and passions, resembling a ghostly puppet show? Or should it be an echoing mausoleum of the vanished, endlessly sweeping out crypts within crypts, akin to a devoted "deconstruction" of the vacated?

However, this apparent dilemma is illusory, as the departure of the objects and the leave-taking in art history are intricately intertwined, forming an inseparable history. The act of taking-leave is not disconnected from what is witnessed departing; it is an ongoing process of mourning the departing. Art history is produced under the "shadow of the object," where the subject witnesses the object's retreat within, creating an interwoven history of art and the art historian. The "shadow of the object" is both the field of death for and in the subject, yet it also serves as an inducement for the ego to live, an existence intertwined with death. Winckelmann's own history exemplifies this intertwining, reconciling his two histories—witnessing and allusion—in his work's ongoing act of mourning. His invention of art history lies precisely in the convergence of these histories, united without complete closure or full restoration. This perpetual mourning sustains art history, making it an enduring entity in division, where the subject and object coexist as a whole.

The intricate entanglement of departure, witnessing, and leave-taking in art history presents an enlightening perspective for discussions on digital necromancy. As technology delves into recreating or simulating the presence of deceased individuals, art history's role as a form of mourning and acknowledgment of loss becomes relevant. Digital necromancy must grapple with ethical considerations, balancing the desire to bring back the departed with the recognition of their eternal absence. Just as Winckelmann's art history confronts the loss of its objects, digital necromancy necessitates acknowledging the irretrievable loss of the past while navigating the complexities of reviving the departed in virtual forms. Embracing this intertwined history of departure and witnessing may offer valuable insights into the implications of digital necromancy and its impact on the delicate balance between preservation and exploitation of the past.

6.3 Fighting Against Loss: The Pet Cemetery Conundrum and the Uncanny Valley

In the realm of digital necromancy, the pursuit to resurrect the deceased echoes a fundamental human endeavor throughout history—to resist loss. Just as Winckelmann's art history grapples with mourning the loss of art's history, the practice of digital necromancy aims to resist the departure of ones lost by recreating their presence in virtual forms. As noted, the urge to maintain a connection with the deceased can be traced to the dawn of human civilization (Ruin 2019). Moreover, the history of necromancy proper reveals a long-standing fascination with the boundary between life and death, where ancient civilizations sought ways to communicate with the deceased, offer guidance, or seek solace in their presence.

Necromancy, as an occult practice, involves the summoning and communication with spirits of the deceased, aiming to gain insights or exert influence from the otherworldly realm. Throughout history, various ancient civilizations and cultures, including ancient Egypt, Greece, and Mesopotamia, have engaged in this mystical art. In ancient Greece, necromancy played a significant role in their beliefs, as they sought to establish connections with the spirit world through elaborate rituals and ceremonies. One prominent example of necromantic practices in Greece was the Necromanteion of Ephyra, a sacred site dating back to the fourth-century B.C.E., where visitors sought prophetic guidance and advice by consulting with the spirits of the dead (Karamanides 2005).

During the Renaissance era, the allure of necromancy continued to captivate the minds of individuals, and one prominent figure associated with such beliefs and practices was the renowned Italian artist and sculptor, Benvenuto Cellini (1500–1571). Despite gaining fame as court artist for the Medici duke Cosimo I, Cellini also delved into the occult and claimed to possess knowledge of necromancy. In his autobiography, *The Life of Benvenuto Cellini* (written between 1558 and 1562), he recounted an extraordinary experience where he attempted to summon a spirit through necromancy. According to Cellini, he performed a ritual in a cemetery, using various incantations and invocations to call forth the spirit of a deceased individual (Gallucci and Gallucci 2003). While the accuracy of the account remains a subject of debate, his involvement in necromantic practices during the Renaissance era highlights the enduring fascination with communicating with the spirit realm.

During the Victorian era (1837–1901), a time characterized by a strong fascination with the supernatural and spiritualism, necromancy took on various forms (Smith 2013). Among them, the notorious case of the Fox sisters, Margaret (1833–1893) and Catherine (Kate) (c. 1839–1892), stands out. These sisters gained prominence as mediums and spiritualists, and in 1848, they claimed to have established contact with the spirit of a deceased individual within their family home in Hydesville, New York. Their alleged ability to communicate with the dead ignited widespread interest in spiritualism, leading to the formation of spiritualist societies and the popularity of séances as a form of entertainment. The purported interactions of the sisters with the spirit realm and their role in popularizing spiritualism during the Victorian era serve as a testament to the enduring allure of necromantic practices across different historical periods (Owen 2004).

The continued fascination with resurrecting the dead in popular culture is evident through various films that delve into the complexities and ethical dilemmas surrounding the grieving process. Movies like *Pet Sematary* (1989), *Flatliners* (1990), and *The Lazarus Effect* (2015) have captivated audiences by exploring the repercussions of bringing deceased loved ones back to life, raising thought-provoking questions about the emotional toll and moral implications of tampering with death.

This fascination reflects a simultaneous attraction and revulsion, akin to the uncanny valley effect. An example can be found in Stephen King's novel (1983) and its film adaptation of *Pet Sematary*, the story revolves around a mystical burial ground with the power to resurrect the deceased. The narrative delves into the consequences of supernatural resurrection when the protagonist, driven by grief, decides to bring back a loved one, only to witness their dark transformation and the devastating effects it has on their family. Similarly, in *Flatliners* and *The Lazarus Effect*, a scientific approach to resurrection is explored. In the latter, a group of researchers discovers a way to revive the dead using an experimental serum. However, the revived individuals return with sinister changes and uncontrollable powers, emphasizing the inherent dangers of meddling with the boundaries of life and death.

Part of the unsettling attraction to such scenarios is the simultaneous promise of reunion (or restitution of what is lost) and revulsion in what that may entail. The experience is similar to the "uncanny valley effect," a term used in robotics, animation, and virtual reality to describe the discomfort or unease experienced by humans when encountering a humanoid or human-like character that closely resembles a real person but still exhibits subtle, unnatural features. The term was coined by the Japanese roboticist Masahiro Mori in 1970 (Mori et al. 2012). According to the concept, as the resemblance to a real human increases, people tend to feel more positive and empathetic toward the character. However, there is a point in the resemblance continuum where the character becomes almost human but not quite, creating a sense of eeriness and discomfort. This dip in the emotional response is the "uncanny valley," and it refers to the phenomenon where human-like entities with slight imperfections or unrealistic features can evoke feelings of revulsion or unease rather than attraction or familiarity. As the character becomes even more realistic and virtually indistinguishable from real humans, the discomfort typically diminishes, and the emotional response becomes positive again.

The uncanny valley effect holds significant importance in domains involving the creation of lifelike human characters, such as robotics, computer graphics, and other related fields. It pertains to how the perception and emotional response of users or viewers can be influenced by the degree of realism in these human-like representations. In the context of digital necromancy, where the aim is to recreate deceased individuals or simulate their presence through advanced technologies like AI and virtual reality, understanding and addressing the uncanny valley effect becomes crucial. Striking the right balance of realism and avoiding the eerie or discomforting aspects can play a pivotal role in how people interact with these digital representations of the departed, impacting their emotional and psychological experiences in significant ways.

6.4 Perpetual Digital Resurrection: From Popular Culture to History

The rapid advancement of technology has ushered in a new era where talking to the dead has become a reality. In popular culture, we have witnessed the emergence of various tools and techniques that enable digital necromancy. Notable examples include hologram concerts and CGI characters that bring deceased individuals back to life in uncanny and captivating ways. One prominent instance of digital necromancy in popular culture is the hologram performance of Tupac Shakur at the Coachella Valley Music and Arts Festival in 2012. Through a clever combination of computer-generated imagery and projection technology, a lifelike holographic representation of the late rapper was brought to the stage, creating a surreal experience for the audience (Moser 2012). Similarly, the world of cinema has also embraced digital necromancy through CGI. In the film Rogue One: A Star Wars Story (2016), the character Grand Moff Tarkin, originally portrayed by Peter Cushing, was digitally recreated to appear in the movie, despite the actor's passing in 1994. The use of CGI to resurrect deceased actors and recreate their performances blurs the boundaries between reality and fiction, leaving audiences both mesmerized and contemplative.

These examples demonstrate how posthumous interaction has become ingrained in popular culture, sparking intriguing questions about the ethical implications and future possibilities of recreating and interacting with deceased individuals through technological means. The concept of conversing with historical icons or walking alongside long-gone figures from the past may no longer be confined to the realm of science fiction. Interacting with realistic clones of historical figures requires several technologies working in tandem to achieve the desired effect. From accurate reconstructions of historical figures and dress to being able to converse with them, and finally being able to be in the same space and physically interact with them are all components that exist and are being combined for such immersive experiences. For instance, YouTuber Denis Shiryaev has harnessed the power of these generative tools to bring famous paintings to life in a unique and fascinating way (https://www.youtube.com/channel/UCD8J_xbbBuGobmw_N5ga3MA). Using neural networks, he has embarked on a venture to generate realistic faces of people depicted in renowned artworks. Among his creations are the living faces of iconic figures from paintings such as Leonardo da Vinci's *Mona Lisa* and *Lady with an Ermine*, Botticelli's model for *The Birth of Venus*, Vermeer's inspiration for *Girl with a Pearl Earring*, and Rembrandt's subjects in *The Night Watch*. Although these recreations are estimations and not authentic historical representations, they provide an example of how innovative AI can be used to create a dynamic connection between the past and the present. It is worth noting that while the generated faces may not always perfectly resemble the originals, they contribute to a novel exploration of historical art and invite audiences to engage with masterpieces in an interactive and imaginative manner.

The integration of AI into the realm of historical exploration and personal connections is further investigated with MyHeritage's AI Time Machine (https://www.myheritage.com/ai-time-machine). As a service for genealogical study, MyHeritage now has a feature that leverages text-to-image AI of Stable Diffusion to transport users into different time periods throughout history once they have uploaded 10–25 of their own photographs in diverse settings and poses. The tool allows users to immerse themselves in the past by transforming their own photographs into hyper-realistic images of historical figures and cast them as the protagonists in multiple predefined themes set in various historical eras (and invented futuristic settings). Users can also share these AI-generated depictions on social media or use them as profile photos (avatars). As part of their suite of advanced photographic technologies, including Deep Nostalgia, Photo Repair, MyHeritage Photo Enhancer, MyHeritage in Color, and DeepStory, AI Time Machine has collectively been utilized over 200 million times since its launch.

Moreover, the applications of AI in the realm of historical recreation and exploration extend beyond photographic transformations. A striking demonstration of the capabilities of XR (Extended Reality) technologies is the meticulous recreation of individual scenarios and even entire civilizations. One such endeavor is the *Virtual Angkor* project undertaken by Monash University in Melbourne, Australia (accessible at https://www.virtualangkor.com/). By harnessing the power of immersive Virtual Reality (VR) and captivating 360-degree videos, Virtual Angkor takes visitors on a mesmerizing journey back to the bustling city during its peak in the thirteenth century. At the heart of Virtual Angkor lies a richly detailed tapestry of 3D models and animated populations that meticulously recreate the historical environment. Through this compelling experience, visitors are immersed in a vivid and historically accurate representation of the ancient city. The project's integration of archaeological findings and textual sources allows for a holistic analysis of various aspects of the civilization, including trade, diplomacy, climate, political power, and the profound sense of place that characterized the medieval complex.

Another interactive worldbuilding example in historical XR experiences is *Witness to Revolution* (2021), a groundbreaking immersive VR game developed in collaboration between the University of Wisconsin-Stout and Carleton College. Accessible at https://www.wttrgame.com/, this immersive game transports players to Colonial Boston, precisely following the events of the infamous 1770 "Boston Massacre." The experience is an ambitious endeavor to present an historically accurate recreation of the period while delving into the profound impact of misinformation on the perception and interpretation of historical events. As players embark on this immersive journey, they are tasked with a crucial mission: to investigate the evidence, interact with witnesses, and experience playable flashbacks of witness accounts. In navigating the complexities of contradictory information, players must confront the challenges of discerning truth from fiction. Throughout the gameplay, players engage with various historical elements, such as Paul Revere's iconic print, "The Bloody Massacre," which serves as a significant artifact of the time. Conversations with individuals and sketching key objects add depth to the experience and enable players to forge their own interpretations of the events that transpired on that fateful day, March 5, 1770.

In 2023, scholars from USC Dornsife College of Letters, Arts, and Sciences are actively engaged in utilizing XR tools to enhance the accessibility of historical objects and places for the public. The Chinatown AR Project (https://dornsife.usc.edu/icw/chinatown-history-project/), for instance, a collaborative effort involving faculty from USC Dornsife and the USC School of Cinematic Arts, is focused on offering visitors to Union Station in Los Angeles an augmented experience of the history of the area. By utilizing archival materials such as photographs and maps, the project aims to create an app that allows users to see a virtual model of the old Chinatown that existed in the early twentieth century. Through this immersive digital recreation, users can gain a sense of the historical context and better understand the impact of its disappearance. Another significant VR project, led by Lisa Pon, a professor of art history at USC, centers around the famous Stanza della Segnatura (a room containing frescos by Raphael) in the Vatican museums. Using VR technology, this interdisciplinary project aims to recreate the environment as it appeared during the reign of Pope Julius II, showcasing his private collection of books. By immersing viewers in this historically accurate setting, the project seeks to offer a deeper appreciation of the relationships between the art, texts, and architecture of that era.

These examples, however, have limited interactive capabilities and the viewer is often treated as a bystander to the action. The recent advancements in generative AI, on the other hand, have introduced exciting possibilities for more interactive and engaging scenarios with historical figures, bridging the gap between the past and the present in a dynamic manner. The Historical Figures Chat app is a notable example of this interactive potential. With the aid of an LLM and neural networks trained on extensive historical text data, the app allows users to engage in simulated dialogues with over 20,000 historical figures. Another app on the market is Hello History-AI Chat app (https://apps.apple.com/us/app/hello-history-ai-chat/id1659654111) where users can choose a historical figure they are interested in and start a conversation, creating a unique and personalized experience. This level of interactivity provides users with a sense of agency, as they actively participate in dialogues with historical personalities, making history feel more alive and relatable. By enabling conversations with both well-known and lesser-known historical figures, these apps offer an opportunity to discover and learn about a diverse range of personalities who have shaped history. Users can delve deeper into the lives, achievements, and impact of these historical figures, gaining insights beyond traditional textbooks and resources. Users can explore historical contexts and engage in historical debates, fostering a deeper understanding of the complexities and nuances of the past. This interactivity not only sparks curiosity but also encourages critical thinking and historical inquiry, making it a potentially valuable educational tool.

There are current limitations, however, such as inaccuracies in the responses and potential whitewashing of certain historical figures. Historians have raised concerns about the need to cross-reference information with verified historical sources to ensure accuracy. In fact, other interactive tools, such as Character AI (https://beta.character.ai/), do not attempt historical veracity at all. The web-based chat offers users the captivating experience of engaging in conversations with historical figures through extensive training on vast text data, the AI system becomes adept at simulating dialogues with a wide range of historical personalities, including iconic figures like Shakespeare, Albert Einstein, Jane Austen, and Socrates, among others. However, all interactions are entirely fabricated, and the AI may occasionally provide banal or inaccurate responses. Despite these challenges, the interactive potential of these apps offer a glimpse into the future of historical interactions, where individuals can engage in meaningful conversations with figures from the past, encouraging a deeper and more personal connection with history. As the AI technology continues to improve and evolve, such interactive experiences may become increasingly accurate, providing an innovative and engaging way to explore history and its profound impact on our world.

Whereas the potential for bringing historical figures to life in a conversational fashion is clear and present, interacting with them in a virtual space, however, has been slow to be developed. There have been examples since 2020, including HistoryMaker VR (https://historymakervr.schellgames.com/), a VR content creation tool designed to enhance history learning experiences for middle school students. The platform offers a unique and interactive approach to history learning by allowing students to step into the shoes of historical figures. With the help of a VR headset (compatible with Oculus Rift/Rift S), learners can immerse themselves in historical settings and embody iconic personalities from US history, including luminaries like Benjamin Franklin, Harriet Tubman, and Sonia Sotomayor. The experience goes beyond mere impersonation, as learners have the opportunity to upload their own speeches and perform them in the virtual environment.

The platform capitalizes on the advantages of VR technology, making history come alive through active physical engagement. Learners can pick up virtual props, gesture, and interact with the virtual environment, enhancing the authenticity of their performances. This level of immersion and interactivity fosters a deeper understanding and connection with historical events and figures. One of the key strengths in examples such as HistoryMaker VR lies in its role as a content creation tool. Learners can bring their creativity to the forefront by customizing their performances with various sets and props. This creative process empowers students to take ownership of their learning, adding a personal touch to their historical experiences.

Another recent example that showcases the potential of VR and augmented reality (AR) in enhancing the visitor experience is the project discussed by Geigel et al. (2020). Their project included the development and implementation of a recreated digital docent, Dr. Frederick F. Backus, original owner of the home, a 3D avatar virtual tour guide at the Genesee Country Village and Museum, the third-largest living history museum in the United States. What sets their approach apart from other museum applications in AR and VR is their focus on narrative storytelling from the perspective of a historical character embodied by a conversational digital avatar. This virtual tour guide, viewed in AR using a HoloLens headset, immerses visitors in a place-based experience within a historical context, featuring a historical figure with a genuine connection to the museum's environment. The project aimed to provide visitors with a comprehensive narrative about life in western New York during the nineteenth century. To achieve this, the team presented five digitally rendered characters and shared their stories through short-form vignettes with the audience. The development team also sought to cultivate "thinking dispositions" in the visitors. By encouraging critical assessment and construction of the significance of historical narratives, the project fosters a deeper engagement with these stories, prompting reflection on a personal, local, and global level. The use of extended reality (XR) in this project presents first-person accounts of the past, blending historical and contemporary elements to create an immersive experience for the visitors. The successful implementation of the virtual tour guide in the context of a living history museum highlights its potential in revolutionizing visitor engagement and historical interpretation.

While the examples provided demonstrate the progress made in bringing immersive experiences and interactivity to historical interactions, there is still much untapped potential in leveraging technology to create even more compelling and authentic encounters with the past. The current state of technology in the fields of XR and generative AI offers exciting possibilities for further advancements in historical recreations and interactive experiences. Innovations in machine learning and improved sensory technologies could allow users not only to see but also to touch, hear, and even smell the past, creating a multisensory historical environment that could revolutionize education and entertainment. Furthermore, as these technologies become more sophisticated and accessible, they hold the promise of enabling users to explore complex historical scenarios and decisions in real time, thereby deepening their understanding of the implications and contexts of historical events.

6.5 Immersive Historical Interactions: Exploring the Limits of Knowing

The advancements in technology are poised to revolutionize historical interactions, ushering in a new era of immersive and engaging experiences. The present state of XR and generative artificial intelligence (GAI) offers promising avenues for further developments in historical recreations and interactive encounters. By harnessing AI-enabled tools, individuals can now create, train, and cultivate digital representations of themselves, ensuring the preservation of their persona to engage in natural conversations even after their passing. This novel paradigm is facilitated by transparent LCD display technology and Light Detection and Ranging (LiDAR), which collectively enable the integration of these digital avatars into real-life settings, giving rise to immersive and authentic interactions with historical figures. The implications of such technological advancements are of substantial academic interest, warranting objective examination and analysis for their potential impact on historical preservation and knowledge dissemination.

Voice emerges as a critical parameter in crafting captivating historical interactions. Voice cloning technology has reached a stage where language models can be trained based on recordings of individuals' voices, giving rise to digital replicas that can partake in meaningful conversations with others. Notably, artists like Grimes have explored the creative potential of AI-generated songs by leveraging language models to create digital clones of their voices. Moreover, the recent introduction of iOS 17 on the iPhone allows users to clone their own voices, making voice replication more accessible than ever before.

Advancements in the creation of digital humans have also brought us closer to virtual selves that bear striking resemblance to our physical appearances. Cutting-edge technologies like Epic Games' Unreal Engine 5 now enable users to scan themselves with smartphones and generate MetaHuman faces tailored to different ages and experiences. This visual aspect complements the experience, facilitating natural interactions with future generations and fostering meaningful conversations about personal life experiences. These advancements hold significant promise for augmenting historical recreations and interactive encounters, elevating the authenticity and engagement of such interactions. However, objective inquiry and academic scrutiny are warranted to assess the implications and potential limitations of these emerging technologies in historical preservation and knowledge dissemination.

Recording experiences emerges as a fundamental aspect in the creation of authentic digital memories. Utilizing tools like the "memo app" or existing communication services, meaningful moments can be captured and securely stored in a designated "memory vault." Projects like Google's Starline further elevate the immersive potential by simulating remote conversations that replicate the feeling of being physically present with others. Project Starline uses advanced hardware and software solutions to capture a three-dimensional representation of users. This 3D representation makes participants appear as if they are sitting across from each other, creating a sense of depth and spatial presence not achievable with traditional 2D video platforms. The technology leverages custom-built depth sensors, which capture a high level of detail, thereby producing hyper-realistic visuals. The resultant image quality surpasses what is currently offered by standard video call platforms like FaceTime. Additionally, the integration of digital humans into physical spaces through semi-transparent LCD displays or holographic devices enhances the realism and interaction capabilities of historical recreations. While verbal interactions remain crucial, tactile experiences also hold significant importance. Emerging technologies such as emerge.io's tabletop device employ mid-air ultrasonic waves to provide users with the sensation of touch and feel for digital objects and interactions. Beyond personal enjoyment, these applications hold immense promise in education and in preserving the wisdom of historical figures such as Einstein, Marie Curie, Gandhi, and Hồ Chí Minh for future generations.

Nonetheless, the concept of digital necromancy raises ethical considerations, particularly regarding the authenticity and impact of reliving digital memories of deceased loved ones. Rigorous testing and thoughtful examination are essential to ensure that these digital interactions enrich and complement real memories rather than undermine or obscure them. As technology continues to advance, the potential for leveraging transparent LCD display technology and LiDAR to perpetuate our digital selves sparks excitement and contemplation. This opens up a realm of possibilities for preserving history and knowledge in unprecedented ways, ultimately shaping how we interact with the past and future in the pursuit of enriched historical experiences. These advancements call for scholarly exploration and objective evaluation to fully understand their implications and limitations in the context of historical recreation and knowledge dissemination.

Despite the significant advancements in AI and related technologies, digital necromancy faces inherent limitations that impact the accuracy and ethical considerations of recreating deceased individuals through AI. Recognizing and understanding these technological constraints is essential for assessing the current capabilities and potential risks associated with this novel approach. One of the primary challenges lies in the accurate representation of individual personalities and experiences. While AI models can learn from vast amounts of data, capturing the nuanced aspects of a person's character, emotions, and subjective experiences remains a complex task. AI systems are likely to struggle in fully comprehending the intricacies of human behavior, motivations, and personal growth, leading to potential inaccuracies and misrepresentations in the recreated digital personas (DiGiovanna 2017). The limitations in replicating the complete essence of a deceased individual's identity raise questions about the authenticity and ethical implications of these digital interactions.

Another limitation involves the potential misrepresentation of cultural context. Cultural norms, values, and beliefs profoundly shape an individual's identity and actions, posing a challenging task to accurately capture and replicate these aspects. AI models trained on diverse datasets may encounter difficulties in fully comprehending the cultural nuances and contextual references that influence a person's behavior, potentially leading to misunderstandings or misinterpretations in the recreated digital representations (Chowdhery et al. 2023). Yet another limitation can be found in the quality and availability of data. The accuracy and fidelity of the recreated personas heavily depend on the data used during training. In cases where limited data is available or the data sources are incomplete, the resulting digital avatars may lack the depth and authenticity required for a faithful representation. As well, cultural norms and beliefs are constantly in flux. Additionally, historical figures or individuals from earlier periods may have limited data available, making it even more challenging to accurately recreate their personas.

In the quest to create accurate digital avatars of historical figures, a diverse array of data sources is required to ensure a faithful representation. These data types encompass photography, audio and video recordings, written records, artistic representations, biographical accounts, and modern scholarly works. Each of these sources contributes valuable insights into the appearance, voice, personality, experiences, and impact on the world that an historical figure may have had. For instance, photography proves indispensable for visual references in the recreation process. High-quality photographs of the historical figure offer essential details for constructing their appearance and facial features. However, the availability of such data depends on the period in which the individual lived. For figures from more recent history, such as the late nineteenth century onwards, photography becomes increasingly accessible after its patenting in 1839, resulting in a higher likelihood of finding photographic documentation. Conversely, for figures from earlier periods, the absence of photographic data poses a challenge in accurately capturing their physical attributes.

Audio and video recordings also provide an invaluable glimpse into the voice, mannerisms, and style of communication of figures. Such recordings could include speeches, interviews, or public appearances. As with photography, the availability of audio and video recordings hinges on the technological advancements of the era. Audio recordings can be traced to the phonograph in 1877 and 1887 for film. Therefore, historical figures of note from the early twentieth century onwards are more likely to have a richer collection of audio and video recordings compared to those from earlier periods. Aside from the direct forms of visual and auditory evidence, there is another category that is relied upon more heavily in figures that predate the nineteenth century. For example, written records, such as historical documents, letters, diaries, and other written materials, furnish significant insights into the historical figure's thoughts, beliefs, and experiences. These records can span centuries, offering a window into the past that helps shape the digital avatar's personality and character. Biographical accounts, autobiographies, and historical works authored by contemporaries or later historians provide crucial information about the individual's life and achievements. The availability and depth of these accounts depend on the historical figure's prominence and the interest in documenting their life during their time and subsequent eras. With regards to the appearance of different individuals, artistic representations, including drawings, prints, paintings, and sculptures, also offer further visual references for recreating historical figures. While such representations can span across various time periods, the level of accuracy and detail in the artworks may vary.

Considering the breadth and depth of data sources, historical figures from the late nineteenth century onwards tend to have more extensive and accurate data available for digital avatar recreation. In contrast, figures from earlier periods, especially before the widespread use of photography and audio recordings, may have significantly fewer data sources, making it challenging to create as accurate and detailed digital representations. As technology and historical research continue to advance, the potential to overcome these limitations becomes an exciting prospect, opening up new possibilities for preserving and engaging with history in immersive and authentic ways. Therefore, while the exploration of historical interactions through immersive technologies and AI-driven advancements has opened up exciting possibilities for engaging with the past in new and dynamic ways, the limitations and a true resurrection of the dead as a simulacra are less likely for more distant figures. The progress made in bringing historical figures to life through AI chatbots, digital avatars, and virtual reality experiences is undeniably impressive. However, as with any cutting-edge technology, it is essential to recognize and address the current limitations that impact the accuracy and ethical considerations of digital necromancy. Understanding the challenges in accurately representing individual personalities, cultural context, and the availability of data is paramount to making responsible decisions in this field. Ethical considerations must be at the forefront of development to ensure that these technologies are applied in a way that respects the privacy, dignity, and historical accuracy of the individuals being recreated.

The knowledge of these limitations can also inform the development of more advanced AI models. By recognizing the complexities of capturing human behavior and experiences, researchers can focus on refining AI algorithms and training methods to achieve more authentic and nuanced representations. Continued research and development in AI and related fields are necessary to address the current challenges and pave the way for a more comprehensive and immersive historical recreation. Moreover, acknowledging the limitations of data availability can help establish realistic boundaries for the application of these technologies. Historical figures from earlier periods may have limited data available, which requires caution in accurately recreating their personas. Responsible practices demand transparency about the sources of data and a clear distinction between historical fact and AI-generated content.

In the pursuit of historical interactions, it is crucial to maintain a balance between technological advancements and ethical considerations. Researchers, developers, and users must collaborate to ensure that digital necromancy respects the rights and memories of individuals while providing enriching and educational experiences. Prior to any digital resurrection, a clear indication of an individual's willingness or reluctance to be virtually cloned is paramount. Integrating such preferences into legal documents, such as wills, offers a concrete mechanism to safeguard an individual's rights posthumously. This proactive approach ensures that the utilization of a person's likeness in virtual realms adheres to their personal desires and ethical considerations, thereby preventing potential misappropriation or unintended emotional distress to surviving kin and acquaintances. By navigating these limitations responsibly and ethically, we can create a future where historical interactions contribute to a deeper understanding of our shared past and the human experience. Continued efforts to address these challenges will pave the way for a more immersive and authentic exploration of history, enriching our understanding of the past and its relevance to the present and future.

References

Chowdhery A, Narang S, Devlin J, Bosma M, Mishra G, Roberts A et al (2023) Palm: scaling language modeling with pathways. J Mach Learn Res 24(240):1–113

Davis W (1994) Winckelmann divided: mourning the death of art history. J Homosex 27(1–2):141–160

DiGiovanna J (2017) 20 artificial identity. In: Robot ethics 2.0: from autonomous cars to artificial intelligence, p 243

Fuller A, Fan Z, Day C, Barlow C (2020) Digital twin: enabling technologies, challenges and open research. IEEE Access 8:108952–108971

Gallucci MA, Gallucci MA (2003) Cellini's poetics II: the vita. In: Benvenuto Cellini: sexuality, masculinity, and artistic identity in renaissance Italy, pp 71–108

Geigel J, Shitut KS, Decker J, Doherty A, Jacobs G (2020) The digital docent: XR storytelling for a living history museum. In: Proceedings of the 26th ACM symposium on virtual reality software and technology, Nov 2020, pp 1–3

Hutson J, Ratican J (2023) Life, death, and AI: exploring digital necromancy in popular culture—ethical considerations, technological limitations, and the pet cemetery conundrum. Metaverse 4(1):12

Karamanides D (2005) Pythagoras: pioneering mathematician and musical theorist of ancient Greece. The Rosen Publishing Group, Inc.

Kaur D, Singh B, Rani S (2023) Cyber security in the metaverse. In: Handbook of research on AI-based technologies and applications in the era of the metaverse. IGI Global, pp 418–435

Mori M, MacDorman KF, Kageki N (2012) The uncanny valley [from the field]. IEEE Robot Autom Mag 19(2):98–100

Morse T (2024) Digital necromancy: users' perceptions of digital afterlife and posthumous communication technologies. Inf Commun Soc 27(2):240–256

Moser JM (2012) Tupac lives! What hologram authors should know about intellectual property law. Bus Law Today 1

Owen A (2004) The darkened room: women, power, and spiritualism in late Victorian England. University of Chicago Press

Rodanés-Vicente JM, Cuchí-Oterino JA, Minami T, Takahashi K, Martín-Gil J, Lorenzo-Lizalde JI, Martín-Ramos P (2023) Use of cinnabar in funerary practices in the Central Pyrenees. Analysis of pigments on bones from the prehistoric burial of the Cueva de la Sierra cave in Campodarbe (Huesca, Spain). J Archaeol Sci Rep 48:103849

Ruin H (2019) Being with the dead: burial, ancestral politics, and the roots of historical consciousness. Stanford University Press

Smith J (2013) The victorian novel and science. The Oxford Handbook of the Victorian Novel, pp 441–458

Winckelmann JJ (1850) The history of ancient art among the Greeks: transl. from the German of John Winckelmann by G. Henry Lodge. Chapman

Xygkou A, Siriaraya P, Covaci A, Prigerson HG, Neimeyer R, Ang CS, She WJ (2023) The "conversation" about loss: understanding how chatbot technology was used in supporting people in grief. In: Proceedings of the 2023 CHI conference on human factors in computing systems, Apr 2023, pp 1–15

Chapter 7
Conclusion

Abstract This chapter reflects on the profound ways in which digital technologies have redefined the landscapes of art, culture, and historical preservation. It emphasizes the transformative potential of extended reality (XR) and other digital tools in fostering more inclusive and interactive engagements with cultural heritage. The narrative reiterates the importance of these technologies in democratizing access to cultural experiences, enabling a broader audience to participate in the global cultural dialogue regardless of geographical and physical limitations. The chapter advocates for a continued exploration of digital possibilities that enhance the understanding and preservation of diverse cultural narratives, promoting a pluralistic approach to art history and cultural studies. It also addresses the challenges and responsibilities that come with the digital representation of culture, including issues of authenticity, ethical considerations, and the potential homogenization of cultural expressions. The conclusion calls for a balanced integration of technology in cultural practices, ensuring that digital advancements contribute positively to the enrichment of human cultural experiences.

7.1 Everyone Speaking at Once (Polylogue): The Fractured Multiverse

The metaverse proper has been hailed as potentially being an interoperable platform that democratizes access to information and collapses time and space. As a virtual shared space, it holds the potential to revolutionize the way we interact, learn, and communicate and offers a vast landscape of immersive experiences, allowing users to explore diverse environments and historical contexts, connecting with others from different parts of the world. The space promises to be a versatile and inclusive one, opening up new avenues for education, research, and collaboration across disciplines. But despite the promise of the metaverse, the current state remains fragmented, with various tech companies controlling different portals and platforms, which limits interoperability and hinders seamless interactions across different metaverse spaces. The fractured nature of the metaverse creates isolated pockets of immersive experiences,

J. Hutson, *Art and Culture in the Multiverse of Metaverses*, Springer Series on Cultural Computing, https://doi.org/10.1007/978-3-031-66320-8_7

restricting the ability to explore and interact beyond individual platforms. This limitation poses challenges for scholars, educators, and creators who seek a cohesive and integrated metaverse experience.

Blockchain technologies or their equivalents have the potential to address the issue of decentralization in the metaverse. By creating decentralized networks and data storage systems, blockchain can enable open access and ownership of virtual assets and experiences that exceed the limitations of industry. The decentralized approach can break down barriers and foster a more equitable metaverse, where information, resources, and opportunities are accessible to all participants. Furthermore, blockchain can enhance security, transparency, and trust in the metaverse, ensuring that interactions are governed by verifiable protocols. As such, in order to achieve the promise of a pluralistic and egalitarian metaverse, it is essential to establish superstructures—the overarching framework, architecture, or infrastructure that governs and supports the virtual shared space—undergirded by blockchain and decentralization (Huynh-The et al. 2023). The framework includes the foundational elements, protocols, and systems that shape the interactions, experiences, and possibilities within the metaverse. The superstructure establishes the rules, standards, and mechanisms that guide how information is accessed, stored, and shared across the various platforms and portals in the metaverse and plays a crucial role in ensuring coherence, interoperability, and a cohesive user experience in the complex and interconnected virtual environment. This foundational structure will foster an inclusive environment where diverse voices and perspectives can thrive. As art and culture play a pivotal role in shaping the metaverse, the centrality of examining their dialogical relationship with commerciality and reproducibility cannot be overlooked. Non-fungible tokens (NFTs) are a prime example of how art and culture are being commodified and reproduced in the metaverse, raising questions about the intersection of art, commerce, and cultural authenticity.

Using Jacques Derrida's "Restitutions of the Truth in Pointing" (1978) as a framing device, the investigation of meaning in the metaverse becomes a nuanced exploration of neoliberalism and life in the fourth dimension (Derrida 1987). The potential of this virtual arena to transcend spatial and temporal limitations raises philosophical questions about the nature of reality, identity, and the boundaries of human experience. By delving into these complexities, we can better understand the transformative power of the metaverse and its implications for society and culture, using Derridian deconstruction to provide insights. There exists a structural affinity between Derrida's deconstruction and the situation of a child discovering the fallibility of its parents while still remaining committed to loving and caring for them. This notion of trust is crucial in addressing the Byzantine Generals Problem in the metaverse, where decentralized networks must navigate uncertainty and conflicting information to achieve consensus (Lamport et al. 2019). The classic problem is a challenge in distributed computing, where consensus needs to be achieved among a group of entities in the presence of faulty or malicious nodes. The problem involves ensuring that all nodes agree on a common course of action despite potential inconsistencies caused by traitors or unreliable communication. Solving this problem is essential for the integrity and reliability of distributed systems, particularly in blockchain and decentralized networks. Consensus algorithms address this issue and enable distributed nodes to reach agreement even with faulty actors. Thus, trust mechanisms rooted in blockchain technologies can provide a foundation for collective decision-making and coordination in the metaverse, ensuring a secure and reliable infrastructure for its pluralistic and egalitarian aspirations.

The same duplicitous considerations are at play in Derridian deconstruction as in the promise of the Byzantine Generals problem and blockchain to resolve issues of trust that ultimately derive from motivations in communication and meaning construction. In fact, the very concept of "deconstruction" according to the poststructuralist philosopher emerged from his critical yet respectful relationship to tradition, where investigating contradictions and exclusions in unity and homogeneity becomes a practice of reckoning. In the context of art history, applying a deconstructive lens raises intriguing implications and calls for an examination of the boundaries and limits of artworks, challenging the notion of their autonomy and considering whether "the visual arts" constitute a distinct code of social meaning production.

Considering one of Derrida's first treatments of the visual arts, *The Truth in Painting (La Vérité en peinture)* (1978), the question arises of whether objects designated as aesthetic can be viewed as relatively autonomous in a semiological sense. The inquiry directly addresses the fundamental concerns of aesthetic philosophers and questions the boundaries and limits of artworks, such as in the Kantian concept of the *parergon*, or the boundaries of a work in connection with this exploration. The further exploration of the "inside" of a work of art reveals that this fundamental question profoundly impacts art discourse, and Derrida argues that both visual and verbal practices are inherently heterogeneous, never existing in a pure or unmediated form. The "inside" of any artwork, he argues, is already inhabited by elements that might have been considered its "outside," such as signatures, verbal discourse, frames, and institutional stagecraft. The notion of the trait or trace, encompassing anything drawn, further blurs the boundaries between writing, speech, and other forms of graphic production.

The investigation that follows is presented as a polylogue, where three interlocutors including Derrida are both juxtaposed and cumulatively superimposed and hold significant relevance in understanding the fragmented nature of art history and its demands for inclusivity and equal representation of all cultures and peoples. The deconstructivist approach of engaging in a polylogue, superimposing three different points of view, offers a pluralistic perspective that acknowledges the complexity and multiplicity of human experiences. By engaging with the philosophical writings of Heidegger and art historian Meyer Schapiro, Derrida deconstructs their arguments to reveal the inherent paradoxes and contradictions within their perspectives on art and representation and challenges the notion of an absolute truth in painting, emphasizing the openness and ambiguity of meaning.

The deconstructive practice of reading exemplifies a pluralistic approach, recognizing that multiple voices and perspectives exist within any discourse. This polylogue is particularly relevant when considering the demands of art history to provide equal voice to all cultures and peoples. Art history as a discipline grapples with the challenge of representing diverse cultural experiences, histories, and artistic expressions. However, attempting to achieve complete inclusivity is virtually impossible due to the immense scope and complexity of human cultural heritage. This impossibility of providing equal representation is also mirrored in the fractured nature of information transference and segmentation in the metaverse. As discussed earlier, the current limitations of the metaverse result from its fractured state, with various platforms controlled by different tech companies, limiting interaction across virtual spaces. The polylogue approach of Derrida aligns with the recognition of the fragmented and diverse nature of information dissemination in the metaverse.

The metaverse, with its potential promise of a unified and democratized platform, presents challenges in reconciling the multiplicity of experiences and voices within it. Just as Derrida critiques the idea of an absolute truth in painting, the metaverse highlights the need for a pluralistic understanding of virtual reality. The fragmented nature of the metaverse, coupled with the challenges of art history, underscores the importance of embracing a diverse and inclusive approach in both understanding and engaging with virtual and physical cultural experiences. As the digital landscape evolves, it becomes imperative to incorporate tools and frameworks that promote accessibility and representation, ensuring that diverse cultural narratives and histories are not only preserved but are also actively explored and appreciated. Additionally, this approach encourages ongoing dialogue and collaboration across different cultural and technological spheres, enhancing the richness of the metaverse as a truly global and culturally comprehensive space.

7.2 Embracing Pluralism: Paving the Way for an Inclusive Metaverse

The metaverse offers immense opportunities for individual and societal advancement, providing immersive and transformative experiences that redefine our interactions with digital content. However, its successful integration and widespread adoption hinge on an in-depth understanding of the diverse and complex factors that influence its evolution. These factors span psychological motivations, social dynamics, technological advancements, educational enhancements, concerns over privacy and security, economic considerations, and the ongoing pursuit of personalization and deep immersion. Recognizing impediments to adoption is crucial for overcoming barriers to access and usability.

As we venture deeper into the metaverse, it becomes crucial to adopt a pluralistic and inclusive approach, acknowledging the segmented and often fragmented transfer of information. Drawing on Derrida's deconstructionist insights to appreciate multiple perspectives, we can foster an equitable and comprehensive virtual space. This approach will not only meet the nuanced demands of art history but also amplify the voices of diverse cultures, thereby creating a more connected and harmonious virtual environment. The critical engagement with our historical narratives is vital for understanding our current societal structures and influencing our future trajectories. As the metaverse expands, we must assess its societal, cultural, and personal impacts thoughtfully. By promoting a pluralistic dialogue and embracing diverse viewpoints, we can navigate the complex demands of art history and the broader digital landscape.

In this era of rapid digital advancement, the metaverse emerges as a crucible for innovation and societal transformation. Attuning to the multifaceted factors that shape its adoption, we aim to cultivate a metaverse characterized by inclusivity, transparency, and ethical integrity. Emphasizing ongoing research, cross-disciplinary cooperation, and a commitment to digital ethics will ensure that the metaverse respects user privacy, ensures security, and achieves high user satisfaction. Looking forward, we must leverage historical wisdom to foster constructive dialogues and approach the evolution of this interconnected digital space with careful consideration and strategic foresight.

References

Derrida J (1987) Restitutions of the truth in pointing. University of Chicago Press

Huynh-The T, Gadekallu TR, Wang W, Yenduri G, Ranaweera P, Pham QV et al (2023) Blockchain for the metaverse: a review. Future Gener Comput Syst 143:401–419

Lamport L, Shostak R, Pease M (2019) The Byzantine generals problem. In: Concurrency: the works of Leslie Lamport, pp 203–226

SPRINGER NATURE

GPSR Compliance

The European Union's (EU) General Product Safety Regulation (GPSR) is a set of rules that requires consumer products to be safe and our obligations to ensure this.

If you have any concerns about our products, you can contact us on ProductSafety@springernature.com

In case Publisher is established outside the EU, the EU authorized representative is:

Springer Nature Customer Service Center GmbH
Europaplatz 3
69115 Heidelberg, Germany

The manufacturer's authorised representative in the EU is Springer Nature Customer Service Centre GmbH, Europaplatz 3, 69115 Heidelberg, Germany. If you have any concerns regarding our products, please contact ProductSafety@springernature.com

Printed and bound by CPI Group (UK) Ltd, Croydon, CR0 4YY
28/09/2026
02224081-0006